Advances in Integrated Pest and Disease Management in Horticultural Crops

Advances in Integrated Pest and Disease Management in Horticultural Crops

Vol. 1

Dr. P. Singh

AGROTECH PRESS

JAIPUR ▪ NEW DELHI

AGROTECH PRESS

Regd. Off. "Bony Residency", Gate No. 2
Opp. Tilak Public School, Vishweswariya Nagar,
Gopalpura Road, Jaipur - 302018, India
Ph: 0141 - 2761280, Telefax: 2761381
E-mail : agrotechpress@gmail.com

Delhi Office: 102, 1st Floor, Satyam House,
4327/3, Ansari Road, Darya Ganj,
New Delhi - 110002, India
Ph: 011-45652440
visit us: www.agrotechpress.com

Distributed By
KEYA INTERNATIONAL PRESS LLP
E-mail: keyainternationalpressllp@gmail.com

Published in India, 2024

ISBN : 9789384568535 (4 Vols. Set)

Typeset by Thomson Press

Printed and bound in India by Thomson Press Ltd.,
New Delhi

Contents

Preface

Pests damage trees, lumber, paper, food and agricultural products in addition to biting humans, livestock and pets, sometimes spreading disease. This book explores the economic, esthetic, medical and nuisance damages caused by pest infestation on fruits and suggest ways to reduce or eliminate further damage using a set of strategies known as Integrated Pest Management. Fruit pests have the potential to cause significant economic losses in fruit production. Even slightly damaged fruit reduces market value. There are a number of lines of defence against pests and diseases in the organic garden, principal among these being the practice of good husbandry, creating healthy soil and ensuring high standards of garden hygiene. But no matter how diverse and healthy the garden eco-system may be, there will always be a degree of disease and pest presence. In many ways, some level of pathogen population in the orchard can be not only acceptable but desirable as they are indicative of a generally healthful and diverse environment, and add to the overall robustness of the system as immunity to such detrimental influences will build up, particularly in a balanced polycultural regime. Good cultural practices—including site selection, variety and rootstock selection, proper planting, pruning, training, fruit thinning, irrigation, and harvest timing—all contribute to a satisfying harvest each year. In particular, well pruned trees allow for increased access to the tree canopy, making pest management tasks such as bagging fruit, monitoring diseases, placing traps, and applying pesticides easier and more successful.

Integrated pest management (IPM) is defined as a comprehensive approach to pest control that uses a combined means to reduce the status of pests to tolerable levels while maintaining a quality environment. Integrated pest management rarely relies on just one tactic—it integrates tactics to prevent pests entirely or reduce them to levels you can live with. IPM is a sustainable approach that allows fruit cultivators to reduce pests to a tolerable level by using the best balance of cultural, physical, biological, & chemical management strategies. "An ounce of prevention is worth a pound of cure," is as true for pest management as it is for human health. Integrated Pest Management is a general strategy of pest damage prevention and control that uses chemical pesticides only when absolutely necessary as a last resort, based on a correct diagnosis and assessment of damage, and only in the safest, most limited way possible. IPM means turning first to cultural practices, pest habitat manipulation, biological controls, mechanical and physical controls (traps and barriers) and the use of resistant varieties. IPM takes into account the level of damage a pest is capable of causing, as well as the possible risks to humans and the environment associated with each pest management strategy. Fruit growers across the world reduce their use of pesticides, while still harvesting quality fruit, through the set of practices of Integrated Pest Management. IPM is based on consistently inspecting and monitoring for pests. The purpose of monitoring is to supply recent, accurate information with which appropriate decisions for managing pests can be made. Guidelines for making appropriate decisions can be established prior to monitoring. Since each site is different, pest management decisions will depend on the circumstances encountered.

The book covers aspects including the ecosystem of an orchard; IPM and fruit-tree physiology and growth; basic biology and management components for insect IPM; basic biology and management components for pests and their natural enemies; economic injury levels for fruit insect pests; tactics for fruit IPM; fruit tree modelling and IPM; biological control models for management of fruit pests; and implementation of IPM. The basic components of an IPM system, including environmental monitoring, pest sampling, pest control models, tactics and implementation are dealt with. The interspecific interactions that take place in a design and management context are described, and how real-time decisions can be made in relation to IPM is shown. The underlying purpose of this book is to formulate a general theory for the evolution of interspecific interactions. The principles of Integrated Pest Management and the applications of these principles for pest management in fruit production are presented. The essential building blocks for the creation of IPM programs are considered; these include surveys of pests, diseases and weeds, alternative practices and non-chemical means of control, monitoring and forecasting methods, and strategies for management and their implementation. The book presents some examples of existing IPM programs in practice. This book provides a holistic view of pest management in fruits, emphasizing the integration of different methods for maintaining pests, diseases and weeds below damaging levels, with the goal of minimizing the use of chemical pesticides that disrupt the environment.

Dr. P.Singh

1

Fruit Industry

Man's relationship with fruiting plants began long before the origins of agriculture in 8000-10,000 BC, when all human beings practised the hunter-gatherer lifestyle. Fruits gathered from the wild were mainstays of our diet, being excellent sources of fibre, vitamins, and other healthful or medicinal compounds unbeknownst to us then. While cereal grains such as wheat and barley were probably the first crop plants domesticated by humans, several of today's fruit crops were not far behind since they were native to the very same area – the fertile crescent of Asia Minor. Domestication of wild fruiting plants may have been inadvertent; the first groves of fruit trees probably sprang from seeds thrown in waste heaps at the edge of villages. Careful observation and selection for useful traits such as larger size, better taste, and higher yield started the transformation of those wild plants into the crops we cultivate and enjoy today. During the age of discovery, fruits, seeds or live plants were often taken on transoceanic voyages, and exchanges in both directions helped spread many crops throughout the world. Christopher Columbus and his contemporaries may not have realized the impact they would have on agriculture and society when they brought crops such as coffee and citrus to the new world, and returned to Europe with previously unknown, but now common foods like cocoa and pineapple.

Today, we have well established world trade networks and sophisticated cultural and postharvest technologies that allow fruits to be enjoyed throughout much of the year, instead of mere weeks per year like our ancestors experienced. Global trade has made formerly rare and exotic treats derived from fruit crops commonplace in countries with no hope of cultivating the plants. Fruit crops are important agricultural commodities, adding tens of billions of dollars per year to the global economy, and being major sources of income for developing countries. Worldwide, over 100 million acres of land has been devoted to their production, and the livelihood of literally millions of farming families depends on continued global trade.

In botany, a fruit is a part of a flowering plant that derives from specific tissues of the flower, one or more ovaries, and in some cases accessory tissues. Fruits are the means by which these plants disseminate seeds. Many of them that bear edible fruits, in particular, have propagated with the movements of humans and animals in a symbiotic relationship as a means for seed dispersal and nutrition, respectively; in fact, humans and many animals have become dependent on fruits as a source of food. Fruits account for a substantial fraction of the world's agricultural output, and some (such as the apple and the pomegranate) have acquired extensive cultural and symbolic meanings.

In common language usage, "fruit" normally means the fleshy seed-associated structures of a plant that are sweet or sour and edible in the raw state, such as apples, oranges, grapes, strawberries, bananas, and lemons. On the other hand, the botanical sense of "fruit" includes many structures that are not commonly called "fruits", such as bean pods, corn kernels, wheat grains, and tomatoes.

The section of a fungus that produces spores is also called a fruiting body.

Botanic Fruit and Culinary Fruit

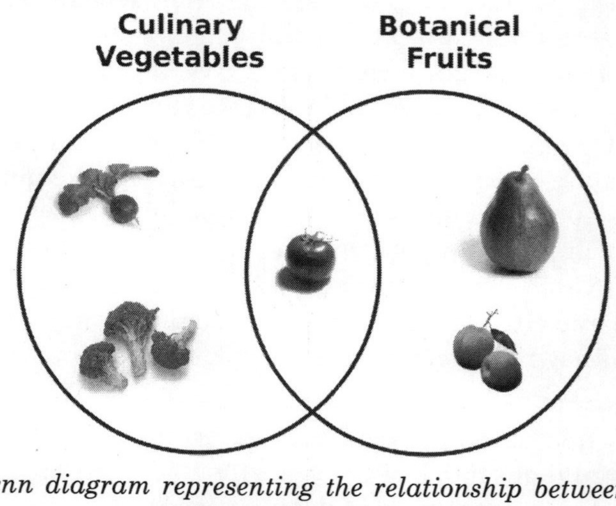

Figure: *Venn diagram representing the relationship between (culinary) vegetables and botanical fruits*

In the culinary sense of these words, a fruit is usually any sweet-tasting plant product, especially those associated with seeds; a vegetable is any savory or less sweet plant product; and a nut is any hard, oily, and shelled plant product. These culinary vegetables that are botanically fruit include cucurbits (e.g., squash, pumpkin, and cucumber), tomatoes, peas, beans, corn, eggplant, and sweet pepper. In addition, some spices, such as allspice and chilies, are fruits, botanically speaking. In contrast, rhubarb is often referred to as a fruit, because it is used to make sweet desserts such as pies, though only the petiole (leaf stalk) of the rhubarb plant is edible. Edible gymnosperm seeds are often given fruit names, e.g., pine nuts, ginkgo nuts.

Botanically, a cereal grain, such as corn, wheat or rice, is also a kind of fruit, termed a caryopsis. However, the fruit wall is very thin, and is fused to the seed coat, so almost all of the edible grain is actually a seed. Many common terms for seeds and fruit do not correspond to the botanical classifications. In botany, seeds are ripened ovules; fruits are the ripened ovaries or carpels that contain the seeds and a nut is a type of fruit and not a seed.

Taxonomy

Plant Names: The scientific or Latin name of a plant is extremely important, as common names vary with location and language spoken. Taxonomic classification places a plant in a large group of related plants (a family), and assigns to it an official name, based largely on Latin and Greek root words. Scientific names generally consist of two italicized words, the first denoting the genus, the second a species within that genus. For example, *Malus domestica* is the name for the cultivated apple, where *Malus* is the genus and *domestica* the species name. In Latin, *Malus* is a noun meaning apple, or alternatively evil, bad or wrong. The species name *domestica* is an adjective meaning "around the house", thus the entire name translates roughly to the domesticated apple. Last but not least, someone always has to take credit for things, so the authority is tacked on to the scientific name denoting the person who named the plant. In the case of the apple, it was a botanist named Borkhausen, so the precise, full name for the apple is:

Malus domestica Borkh: Linneaus, the father of botany, named many plants, which explains the capital letter "L" found at the end of many plant names, such as *Pyrus communis* L. (pear). The authority is often omitted in popular literature, on nursery tags, or in plant catalogues, so outside of scientific literature it is not often seen. The authority is not italicized, but typically abbreviated, and there are literally thousands of these abbreviations in use today.

To make matters more complex, some plants have been named or renamed several times, so two or more scientific names may be found for the same plant. For example, the Asian pear, *Pyrus pyrifolia* (Burm. f.) Nak., also is named *Pyrus serotina* L., so you may see either name used. Yes, this somewhat defeats the purpose of using a unique, scientific name for a given plant, but opinions vary as to which is "right".

We are all familiar with different types of fruits within a species, such as 'Red Delicious', 'Golden Delicious', and 'Granny Smith' apples. Horticulturists refer to these subspecies as "cultivars", which means "cultivated varieties". The term "variety" often is used interchangeably with cultivar, although purists prefer the latter. Note the convention of enclosing the cultivar name in single quotes.

An even finer level of detail is found in some plants, the strain or sport. This is equivalent to a sub-subspecies, or a form, where within a cultivar we have several slight variations that have horticultural importance. An example would be 'Ultra Earli Fuji' apple, which is a strain of 'Fuji' apple that ripens a bit earlier than the original 'Fuji'. As in this example, strains are often given a new cultivar name if they become commercially important. Another example is 'Red Max', a strain of 'McIntosh' apple with deeper red colour. While this seems like the splitting of hairs, it is useful to note that, in this example, 'Red Max' and 'McIntosh' are genetically more similar than are 'Fuji' and 'McIntosh'. A few more points will help clarify the use of plant names:

- After the first reference to a genus in the text, the genus name is often abbreviated using only the first letter. For example, in the chapter on apple, I discuss species related to *Malus domestica* and refer to these as *M. floribunda, M. sargentii, M. micromalus,* etc., hoping that you understand that the "*M.*" stands for "*Malus*" in each case; this saves valuable typing time for the author and a few drops of ink for publishers.
- A multiplication sign "*X*" placed between the genus and species denotes an interspecific hybrid within that genus. An example is the name for the cultivated strawberry, *Fragaria X ananassa,* which was originally developed by crossing *Fragaria chiloensis* with *Fragaria virginiana.*
- Some genera of plants are so poorly characterized, so diverse, or contain hybrids with complex parentage derived from so many species, that it is convenient to refer to them as a group using the abbreviation "*spp.*". Among fruit crops, blackberries are a good example of this, often referred to as *Rubus spp.*

Cultivars

For each crop, I discuss only the major cultivars or groups of cultivars. Even brief descriptions of the many cultivars grown would require a doubling of the length of the text. Cultivars change over time, and vary across different fruit growing regions, making it impossible to cover adequately and keep the text as concise as possible. Several other texts or sources are devoted to cultivar descriptions, including some listed below, and I have included one or two references for each crop that give more detail on cultivars.

Origin, History of Cultivation

The centre of diversity of a plant species denotes the area of the world where the species evolved and is found growing in the wild. It is interesting to note that many of the fruit crops (and crop plants in general) grown in the United States are not native to North America. Perhaps more interesting is that several cultivated fruits are not found in the wild, indicating that man has hybridized or selected the species over time to make it very different from its wild progenitors. It is not surprising, therefore, that many of the world's most common fruits are native to the area where agriculture had its roots – Asia Minor and China.

Knowledge of soil and climatic conditions occurring in the native range give us clues about site selection and cultural methods that should be employed when growing these crops in foreign areas. Also, plant breeders often return to these centres of diversity to collect germplasm useful for disease and pest resistance or other traits. A brief history of cultivation is provided for each crop in the text. Lessons from history have been valuable in shaping the way we grow crops today.

Folklore, Medicinal Properties, Non-food Usage

Fruits have more than just nutritional value to us; some may contain anti-cancer compounds while others may cause health problems or even death. In this section, I have compiled information from various sources on the folklore, myths, healing properties, and symbolism that has surrounded fruits for centuries. Some of the information on medicinal properties is well-documented, but some of it is speculative.

Most fruit crops are perennial trees, shrubs, or vines. Trees are large woody plants which generally produce a single main stem or trunk, where the renewal growth occurs at the shoot tips in the canopy. The latter is an important distinction between trees and shrubs, since large shrubs can be trained to a single stem, but tend to produce new growth from the base or crown.

Vines or lianas are woody plants that are trained to have a single trunk at the base, but use twining stems or tendrils to support the canopy. Vines rarely have large trunks like trees since they support themselves by climbing on taller plants in nature, or on trellises in cultivation. As a result, vines spend little of their energy on supportive wood, while growing very tall and maximizing leaf exposure to sunlight.

Leaves take many forms, being compound if composed of two or more leaflets, or simple if just a single leaf blade.

Fruit Tree

A fruit tree is a tree which bears fruit that is consumed or used by humans and some animals — all trees that are flowering plants produce fruit, which are the ripened ovaries of flowers containing one or more seeds. In horticultural usage, the term 'fruit tree' is limited to those that provide fruit for human food. Types of fruits are described and defined elsewhere, but would include "fruit" in a culinary sense, as well as some nut-bearing trees, such as walnuts.

The scientific study and the cultivation of fruits is called pomology, which divides fruits into groups based on plant morphology and anatomy. Some of those groups are: Pome fruits, which include apples and pears, and stone fruits, which include peaches/nectarines, almonds, apricots, plums and cherries.

Fruit Tree Forms

Fruit trees are grown in a variety of shapes, sometimes to please the eye but mainly to encourage fruit production. The form or shape of fruit trees can be manipulated by pruning and training. Shaping and promoting a particular tree form is done to establish the plant in a particular situation under certain environmental conditions, to increase fruit yield, and to enhance fruit quality.

For example, pruning a tree to a pyramid shape enables trees to be planted closer together. An open bowl or cup form helps sunlight penetrate the canopy, thus encouraging a high fruit yield whilst keeping the tree short and accessible for harvesting. Other shapes such as cordons, espaliers and fans offer opportunities for growing trees two dimensionally against walls or fences, or they can be trained to function as barriers.

Forms

Some of the following fruit tree forms require training by tying the branches to the required form. Most require pruning to retain the desired structure. However, not all types of fruit tree are suitable for all forms; apples and pears do well as cordons and espaliers, for example, whereas cherries are more suited to the fan form.

Fruit tree shapes

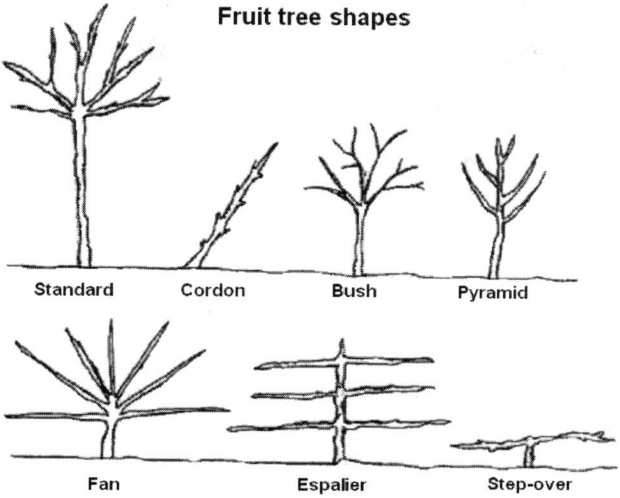

Bush

An open-centred crown on a short trunk of less than 1 metre (3 ft 3 in). This is a traditional and popular form for apple trees. Bush trees are easy to maintain and bear fruit at a young age. Final height is between 2 metres (6 ft 7 in) and 5.5 metres (18 ft), depending on which rootstock is used.

Standard

Larger than the bush form, with trunks of 2 metres (6 ft 7 in) or more. Standard trees can reach a total height of 8 metres (26 ft). They eventually produce high yields but, being large trees, are not easy to maintain.

Pyramid

Similar to the bush form, although the main leader shoot is allowed to maintain its dominance, resulting in a pyramidal shape.

Spindlebush

A variant of the pyramid form in which the lateral branches are tied down to a horizontal position. Designed for dense orchards by Schnitz-Hubsh and Heinrichs in Germany in 1936, this is currently the most popular training system for dwarf apple and pear trees.

Cordon

Single-stemmed trees planted at an angle (usually 45°), with fruiting spurs encouraged to form along the stem. Any side branches are removed by pruning. Cordons take less space and crop earlier than most other forms, so more varieties can be got into a small space, but yields are smaller per tree. A special cordon set-up is the Bouché-Thomas system.

Espalier

A central vertical trunk with three or four horizontal branches on each side. A special espalier in this group is the LePage-system.

Fan

A short central trunk with several radiating branches growing from the crown.

Step-over Espalier

Espaliers with just one tier of horizontal branches 30 cm from the ground. These make a novel and productive border for a vegetable plot.

A study on orchard mango trees in Nelspruit, South Africa, compared open vase, closed vase, central leader, palmette and standard pruning systems and recommended a modified pyramid, somewhere between a central leader and a closed vase system, for high-density mango orchards. The study also evaluated both post-fruit-set and post-harvest pruning, indicating that late mango cultivars benefit from pruning while bearing in late fall, while early cultivars may be best pruned immediately after harvest.

Yield and Spacing

Apples and pears	*Yield*		*Spacing*	
	Apples	*Pears*	*In rows*	*Rows apart*
Bush	25–50 kg	20–45 kg	4–5 m	4–5 m
Dwarf bush	15–25 kg	10–20 kg	2.5–5 m	2.5–5 m
Dwarf pyramid	5–7 kg	3–5 kg	1.5–2 m	2 m
Espalier (two tier)	10–12 kg	7–10 kg	3–6 m	2 m
Fan	5–15 kg	5–15 kg	4–5 m	-
Single cordon	2–4 kg	2–3 kg	0.5–1 m	2 m
Standard	50–200 kg	40–100 kg	6–10 m	6–10 m

Other tree fruits	*Yield*	*Spacing*	
		In rows	*Rows apart*
Bush (sour cherry)	15–20 kg	4–5 m	4–5 m
Bush (plum and peach)	15–30 kg	4–5 m	4–5 m
Bush, standard (sweet cherry)	15–50 kg	5–12 m	5–12 m

Contd...

Other tree fruits	***Yield***	***Spacing***	
		In rows	***Rows apart***
Fan (all stone fruits)	7–15 kg	4–5 m	-
Fan (sweet cherry)	6–15 kg	5-7.5 m	-
Pyramid (plum)	15–25 kg	3–4 m	3–4 m
Standard (plum, peach and apricot)	15–50 kg	5-7.5 m	5-7.5 m

Orchard Planning, Layout and Management

An orchard is an area which is dedicated to the growing of fruit or nut trees and sometimes shrubs which produce fruit and nuts. Orchards vary widely in layout, shape, and size, depending on regional preferences and the types of trees involved, and many people have fond associations with orchards, as they produce favoured treats like peaches, cherries, apricots, almonds, and many others. Even a small orchard can be amazingly productive, which is why some homes in rural areas have a planting of a few fruit trees to supply household needs.

Humans have been establishing orchards for a very long time, and until the 15th century, the word was also used to refer to vegetable gardens. Pioneers in the world of agriculture undoubtedly learned a great deal through the domestication of fruit trees, finding out how to breed desirable varieties, making grafts and slips, and learning about the best conditions for growing various types of trees. The remains of ancient orchards can be seen in many regions of the world, and some orchards have trees which are hundreds of years old and still producing.

The primary feature in an orchard is obviously fruit and or nut trees. Some orchards also have a cover crop which can range from flowers to vegetables, keeping the soil well conditioned and making the orchard more productive. In other regions, people allow their animals to graze in their orchards, keeping the grass in the orchard clipped, providing fertilizer for the trees, and disposing of windfall fruits. Many orchards are fenced to keep out pests, and netting may be used to keep birds away from the fruit as it grows. Beekeepers are fond of keeping their bees in an orchard, taking advantage of the numerous flowers which appear in the spring.

Planning an orchard requires care. When mixed trees are planted, gardeners need to ensure that the soil and weather conditions are right for all the trees, and the trees must be well spaced to allow them to grow. The orchard must also receive regular care to ensure that it produces, with people pruning the trees annually to shape them and promote healthy production. Gardeners also need to consider how the trees will exchange pollen for fertilization, in the case of trees which are not self-pollinating.

Many orchards are planted in a grid format, making it easy to work and harvest in the orchard. However, orchards can also be more naturally scattered, to take advantage of prevailing soil and weather conditions, and these natural orchards may be made ornamental as well as productive with cover crops of flowers and benches which allow people to sit in the orchard, enjoying the presence of fruit trees and flowers.

In some regions, it is possible to visit an orchard and pick your own fruit. Several botanical organizations also maintain orchards of heritage fruit trees to ensure that unusual cultivars do not die out. Many of these organizations are happy to send grafts to interested gardeners, keeping rare fruit and nut varieties out and about for future generations to enjoy.

Classification of Fruit Crops

With hundreds of thousands of plants used by mankind, it is impossible to talk about each one individually. Plants are grouped by various common characteristics to help us communicate similar ecological adaptations and cultural requirements. For example, the term "shade plants" indicates plants tolerant to various levels of shade. "Xeric" groups those plants requiring less supplemental irrigation in our climate. It is important to point out that any classification system will have plants that don't exactly fit the groupings.

From time to time, horticultural crops have been classified into various groups depending on their growth habits, cultivation requirements, climatic needs and uses. Horticultural crops are popularly classified into the 3 broad divisions of fruits vegetables and flowers.

However, following points are considered to be most important basis for classification of the crops.

1. Classification Based on Growth Habit and Physiological Character.
2. Classification Based on Life Span of Plants.
3. Classification Based on Climatic Requirements.
4. Classification Based on Seasonal Basis.
5. Classification Based on Use of Horticultural Plants.

The following are examples of some common classifications used in horticulture.

Classification by Use

I. Edibles
   A. Fruits
      1) Tree fruits
      2) Small fruits
   B. Vegetables
      1) Warm season vegetables
      2) Cool season vegetables
   C. Herbs
      1) Culinary
      2) Medicinal
   D. Nuts
II Ornamentals/Landscape Plants
   A. Woody plants
      1) Trees
      2) Shrubs
      3) Vines and ground covers
   B. Herbaceous plants
      1) Flowers
      2) Vines and ground covers
   C. Grass/turf
III. Potted plants, houseplants, gift plants
   A. Flowering gift plants
   B. Foliage plants

Note: Do not confuse the multiple uses of the word "fruit". In reference to "fruits and vegetables", "fruit" refers to crops primarily used in some European cuisine as a dessert (peaches, apples, strawberries, and raspberries). "Vegetables" refers to crops served as part of the main entrée (potatoes, carrots, corn, and lettuce). In this frame of reference, tomatoes are vegetables. In reference to "fruit" as a part of plant anatomy (i.e., roots, stems, flowers, fruits and seeds) tomatoes, squash and watermelons are fruit.

Classification by Climatic Requirements

Temperature Requirements

- Tropical plants originate in tropical climates with a year-round summer-like growing season without freezing temperatures.

Examples include cocao, cashew and macadamia nuts, banana, mango, papaya, and pineapple.

- Sub-tropical plants cannot tolerate severe winter temperatures but need some winter chilling. Examples include citrus, dates, figs, and olives.
- Temperate-zone plants require a cold winter season as well as a summer growing season, and are adapted to survive temperatures considerably below freezing. Examples include apples, cherries, peaches, maples, cottonwoods, and aspen. In temperate-zones, tropical and sub-tropical plants are grown as annuals and houseplants.
- Cool season plants thrive in cool temperatures (40°F to 70°F daytime temperatures) and are somewhat tolerant of light frosts. Examples include Kentucky bluegrass, peas, lettuce, and pansies.
- Warm season plants thrive in warm temperatures (65°F to 90°F daytime temperatures) and are intolerant of cool temperatures. Examples include corn, tomatoes, and squash. Some warm season plants are sub-tropical and tropical plants grown as annuals in Colorado.
- Tender plants are intolerant of cool temperatures, frost, and cold winds. (e.g., most summer annuals, including impatiens, squash, and tomatoes).
- Hardy plants are tolerant of cool temperatures, light frost, and cold winds (e.g. spring-flowering bulbs, spring-flowering perennials, peas, lettuce, cole crops).
- Hardiness refers to a plant's tolerance to winter climatic conditions. Factors that influence hardiness include minimum temperature, recent temperature patterns, water supply, wind and sun exposure, genetic makeup, and carbohydrate reserves.
- Cold hardiness zone refers to the average annual minimum temperature for a geographic area. Temperature is only one factor that influences a plant's winter hardiness. As of printing time, a new U.S. Department of Agriculture (USDA) Hardiness Zone Map is ready for release. With an extensive database, it will clear up some of the confusion created in previous versions and documents some zone creeping (zone moving north as temperatures warm).
- Heat zone refers to the accumulation of heat, a primary factor in how fast crops grow and what crops are suitable for any given

area. This is only one factor that influences a plant's heat tolerance. On a Heat-Zone map, the Colorado Front Range falls into zones 5 to 7.

Classification by Elevation and Plant Life Zones

Higher elevations have increasingly shorter growing seasons due to colder temperatures. High elevations have drier soils, stronger light, persistent winds, and greater temperature changes. Due to this harsh environment, alpine and tundra plants tend to be compact in form.

Classification by Ecological Adaptations

Many of our plant care problems arise as gardeners try to grow plants outside of their natural environment or "ecological adaptation."

Characteristics of the Colorado high plains include low humidity, limited rainfall, and alkali soils low in organic matter. One of our most limiting factors is rapid temperature change. We can go from a beautiful warm spring day to a cold winter blizzard in just hours.

In our higher mountain communities, the short frost-free season and low summer growing temperatures significantly limit plant selection.

The following are a few examples of terms used to describe classifications based on ecological adaptation.

- Alpine plants tolerate the short growing season, cold, and wind of higher mountain elevations. They are typically low-growing, small leaf perennials. Snow cover depth often dictates the plant's growing height.
- Prairie plants are adapted to the open sun and winds of the plains. These plants are further classified into dry, mesic, and wet prairie categories.
- Woodland plants are adapted to a low light conditions and soils rich in organic matter. They typically have large leaves and small flowers.
- Wetland plants tolerate continually moist soil conditions of a bog or a pond. Wetlands play a primary role in water quality as a filtering system for water-borne pollutants.
- Xeric plants tolerate conditions of low water, bright light, and warm temperatures due to a variety of adaptations such as thick, waxy, or fleshy leaves, hairy leaves, small narrow leaves, taproots and succulent stems. An excellent text on xeriscape gardening is *Xeriscape Plant Guide,* by Denver Water, published by Fulcrum Publishing.

Native and Adapted Plants for the Urban Environment

- Native (indigenous) plant refers to plants adapted to a given area during a defined time period. In America, the term often refers to plants growing in a region prior to the time of settlement by people of European descent.

The term is so overused that it has little meaning. With recent interest in water conservation, many gardeners mistakenly consider "native" plants as "xeric" plants, and "xeric" plants as "native" plants. The two terms are not interchangeable.

The concept of native should not refer to political boundaries, such as state or country, but rather to an ecological habitat during a defined chronological period. For example, Colorado blue spruce and quaking aspen are "native" to the ecological habitat referred to as the montane zone. They are not "native" to the Colorado high plains, or elevations below 8,000 ft. From a chronological reference point, what is now the grassland of the Great Plains was once an inland sea. Therefore, aquatic plants such as kelp would have been "native" at one time. Over time, the ecological habitat changed, changing the "native" plants along with it. Environmental change is an ongoing process, based both on global climatic events and on the activity of all organisms, including mankind.

- Adapted or introduced) plants are those that reliably grow well in a given habitat without specific attention from humans in the form of winter protection, soil amendments, pest protection, water, etc. Adapted plants may be considered to be *low maintenance* plants.
- Urban environment – For gardening purposes, the urban setting needs to be recognized as a unique ecosystem. Characteristics of the urban environment include:
- Soil compaction
- Rooting areas covered with buildings, roads, and parking lots
- Increased surface runoff creating significant water quality problems
- Higher temperatures and lower humidity
- Air pollution

Characteristics of an urban environment cultivated by humans may include:

- Reduced wind
- Increased availability of water due to irrigation

- Increased organic matter and soil fertility
- Reduced pests
- Increased soil stability
- Slower temperature fluctuations

Classification by Stem and Leaf Texture

- Herbaceous plants have non-woody stems.
- Woody plants have woody stems that generally live for several years, adding new growth each year.
- Deciduous plants shed all leaves at approximately the same time annually.
- Evergreen plants retain some leaves longer than one growing season so that leaves are present throughout the year. Seasonal drop of some of the oldest interior leaves is a natural part of the life cycle.
- Semi-evergreen refers to plants that may retain their leaves, depending on the winter temperature and moisture.
- Broadleaf plants have a broad leaf blade (e.g. ash, maple, lilac and beans).
- Narrowleaf plants have needle-like (e.g. pine, spruce) or awl-like (e.g. junipers) leaves.
- Grass-like plants have narrow leaves, usually arising from the base of the plant. The leaves may be soft (ornamental grasses) or stiff (yucca).

Reminder:

- Some evergreens are broadleaf (e.g., Oregon grape, most true hollies, and evergreen euonymus).
- Some narrow-leaf plants are deciduous (e.g., larch and bald cypress).
- *Conifer* refers to cone-bearing. Most conifers are narrow-leaf evergreens. A few conifers are deciduous (larch, bald cypress).

Classification by Growth Habit

Growth habit refers to the genetic tendency of a plant to grow in a certain shape and to attain a certain mature height and spread.

- Trees typically have a single trunk and mature height over 12 feet.
- Shrubs typically have multiple-branches from the ground and a mature height less than 12 feet.
- Vines have a climbing, clasping, or self-clinging growth habit.

Note: Many landscape plants could be considered small trees or large shrubs. The terms "tree" or "shrub" would be applied based on the general appearance of the plant.

Plants have vastly different growth habits. It is important to understand growth habits in order to make knowledgeable decisions regarding plant placement, plant selection, pruning and maintenance requirements.

The species, cultivar, and/or variety name sometimes indicates some characteristic of growth habit.

Figure: *Plant forms.*

Classification by Life Span

From a horticultural perspective, life span is a function of climate and usage. Many garden plants (including tomatoes and geraniums) grown as annuals in Colorado would be perennials in climates without freezing winter temperatures.

- Annuals complete their life cycle (from seedling to setting seed) within a SINGLE growing season. However, the growing season may be from fall to summer, not just from spring to fall. These plants come back from seeds only.
- Summer annuals germinate from seed in the spring and complete flowering and seed production by fall, followed by plant death, usually due to cold temperatures. Their growing season is from spring to fall. Examples include marigolds, squash, and crabgrass. These are also called warm season annuals.
- Winter annuals germinate from seed in the fall, with flowering and seed development the following spring, followed by plant death. Their growing season is from fall to summer. Examples include winter wheat and annual bluegrass. These are also referred to as cool season annuals. Many weeds in the lawn (such as chickweed and annual bluegrass) are winter annuals.
- Biennials complete their life cycle within TWO growing seasons. Biennials germinate from seed during the growing season and often produce an over-wintering storage root or bulb the first

summer. Quite often they maintain a rosette growth habit the first season, meaning that all the leaves are basal. They flower and develop seeds the second summer, followed by death.

In the garden setting, we grow many biennials as annuals (e.g., carrots, onions, and beets) because we are more interested in the root than the bloom. Some biennial flowers may be grown as short-lived perennials (e.g., hollyhocks).

- Perennials live through several growing seasons, and can survive a period of dormancy between growing seasons. These plants regenerate from root systems or protected buds, in addition to seeds.
- Herbaceous perennials develop over-wintering woody tissue only at the base of shoots (e.g., peony and hosta) or have underground storage structures from which new stems are produced. (Please note: Golden Vicary Privet can be either herbaceous or woody as grown in Colorado.)
- Spring ephemerals have a relatively short growing season but return next season from underground storage organs (e.g., bleeding heart, daffodils).
- Woody perennials develop over-wintering tissue along woody stems and in buds (e.g., most trees and shrubs grown in Colorado).
- Combinations – Plants are usually classified as annual, biennial, or perennial on the basis of the plant part that lives the longest. For example, raspberries have biennial canes and perennial roots.

2

Tropical Fruits

Tropical fruits having remarkably great qualities, you can always eat them unprocessed and without any added sugars. Being an important source of carbohydrate, vitamins, minerals and fibres, tropical fruits grow on plants of all habitats. These fruits are a primary source of nutrition and a delicious component of healthy, balanced diet. Across the world, there are different types of tropical fruits grown and exported.

Abiu

Pouteria caimito, the abiu, is a tropical fruit tree originated in the Amazonian region of South America. It will grow an average of 33 feet (10 m) high, and can grow as high as 116 feet (35 m) under good conditions. Its fruits shape varies from round to oval with a point. When ripe, it has smooth bright yellow skin and will have one to four ovate seeds. The inside of the fruit is translucent and white. It has a creamy and jelly-like texture and its taste is similar to the sapodilla — a sweet caramel custard. The abiu tree is part of the Sapotaceae family and is very similar in appearance to the canistel.

Distribution

The abiu, *Pouteria caimito*, is commonly considered as native to the headwaters of the Amazon. It grows wild in the lower eastern part of the Andes from southwestern Venezuela to Peru. It also grows around Tingo Maria and Iquitos, Peru and it will commonly be found in the Province of Guayas in Ecuador, where it's sold in the markets. Due to early and widespread cultivation by Amerindians, its original distribution in Brazil, outside the Amazon, is uncertain. In the Amazon basin, it is found to grow heavily in Pará but is also found sparsely in collections from the Atlantic rainforest near Rio de Janeiro and Bahia. It can also be found in Colombia in areas such as the regions of Caquetá, Meta and Vaupes and it is very plentiful in Amazonas, Venezuela. It has also been growing for a very long time in Trinidad.

The abiu grows best in areas that have a year-round moist and a warm climate. It can now be found throughout most of the Amazon. It is a common dooryard tree in the backyards and streets in the city of many Brazilian towns, but it is not usually grown commercially. The abiu habitats are nearly all tropical. It will thrive in a place that has a year round warm and moist climate, although it has been known to grow well in Rio de Janeiro, which is a somewhat cooler climate. In Peru it cannot grow above 2,000 feet (610 m) feet in elevation, but in Colombia it has been found growing up to an elevation of 6,000 feet (1,800 m).

Other Names

The fruit is known by different regional names in countries of production: in Trinidad it is the yellow star apple or caimitt (distinct from the star apple); Colombians know it as the *camio*, *caimito amarillo* (again not to be confused with *Chrysophyllum cainito*, which is known as *caimito* in some countries) or *madura verde*; in Ecuador it is known as the *luma* or *cauje*, while in Venezuela it is known as the *temare*. Other common names are *abio*, and (in Portugal) *abieiro*.

Description

The leaves of *Pouteria caimito* range from oblong to elliptic. The can be anywhere from four to eight inches (10 to 20 centimetres) in length and one and a half to two and a half inches (3.5 to 6.5 centimetres) in width. The flowers on the tree can come either by themselves or in clusters of two to five flowers. They will appear on the leaf axils on long, thin shoots. The flowers are small with four to five petals. The petals are cylindrical and are white to greenish in colour. The flowers are hermaphroditic, meaning they are both sexes. The flowers open in the morning and can stay open for about two days.

Fruit

Mature abiu trees produce one hundred to one thousand fruits each year. These have a pale, translucent pulp of a custard consistency that is easily scooped out with a spoon; there may also be a few bits of tougher gel. The seeds are easily removed and are covered with a thin layer of adherent pulp. The fruit has a sweet, mild taste which may have a hint of pineapple but is best described as reminiscent of caramel flan. It is often used in ice cream or eaten out of hand.

Unripe fruits contain a gummy and unpalatable latex that hardens upon exposure to air. The skin of the ripe fruit is a pale yellow colour with a leathery texture and residual latex. Because mature fruits will continue to ripen when picked, the harvest can be timed to allow for transportation to market. However, this period may be as short as five days. Maturation can be recognized by the pale green-to-yellow colour break and the ripe fruit can be identified by its yellow colouration and a slight softness.

Varieties

The abiu varies a lot in form, size and quality of the fruit. Some have soft and others firm flesh. Some taste bland while others have very nice flavour. In Colombia along the Putamayo River there is a variety with large, round fruits that produces after four years. In Vaupes, Colombia, there is a variety that will bear fruit just one year after seeding; however, these fruits will be small and contain little pulp.

Season

Abiu may have several flowering periods a year, with potential for both flowers and fruit on the tree at one time. The development time from flower to ripe fruit is approximately three months. The main crop season varies by climate.

In Ecuador, the fruits are in season during the months of March and April, but may also produce in other times of the year. In some Brazilian markets, they are sold from September to April, however at this time they are sparse. In Bahia there is a short season in February and March. In the small region of suitable subtropical climate in the continental United States, the southern tip of Florida, the fruits ripen from August to October. In Northern Queensland, Australia, the main crop ripens January to March.

Uses

The wood of the abiu tree is dense, heavy, and hard and is used as lumber in construction.

Culinary

The fruit of the abiu tree is edible and considered one of the best of the sapotes due to having the sweet caramel-like taste of sapodilla with a smoother texture. It is commonly eaten out of hand and, although in Colombia those eating the fruit this way are advised to grease their lips to keep the gummy latex from sticking, this hazard can be avoided by selecting fully ripe fruits and scooping out the flesh with a utensil. The tartness of a bit of added lime juice may enhance the flavour, especially when chilled.

The melting sweet pulp of the abiu is also used to flavour ice cream and cut into yogurt for a light and delicious breakfast. The subtlety of the flavour limits its utility in more complex confections and salads. Abiu fruit is a significant source of calcium, phosphorus, vitamin A, and vitamin C.

Medicinal

The dark golden-yellow, overripe fruit develops a mucilaginous nature. In Brazil, people use this characteristic to relieve coughs, bronchitis and other pulmonary afflictions. Other folk medicine uses are as an astringent, anti-anemic, and anti-inflammatory and to relieve fever and diarrhea.

Cultivation

The tree grows best in tropical areas and in places that have a warm moist climate all year long, and is slightly less hardy than related sapotes such as the canistel and sapodilla (nispero). Within the United States, it grows well in South Florida as far north as Palm Beach County, and has survived brief freezes. It prefers wet, slightly acidic soil with high organic content and may suffer from iron deficiency (chlorosis) in alkaline soils. A tree that has just been planted will be fragile and needs protection from wind and cold weather. Only light pruning is necessary and feeding should be frequent, but light.

Propagation

Propagation is nearly always by seeds, but the fruit of seedling trees is variable. Fresh seeds need to be planted within a few days if they are to remain viable and will germinate in two to three weeks. The lower branches of the seedling may be pruned after a year and the first fruit can be expected in three years, with substantial yields at five years. Grafting, budding and air layering can be used to propagate superior strains and advance the production schedule.

Pests and Diseases

There is little commercial value to the fruit in its native range because it is often damaged by small insects. In Brazil, the main pest that destroys the fruit is thought to be the fruit fly. Thrips are known to attack buds, leaves and young fruits, but in exotic locations abiu can be relatively pest-free, especially if fruit is harvested at maturity and ripened off the tree.

Sugar Apple

Sugar-apple is the fruit of *Annona squamosa*, the most widely grown species of *Annona* and a native of the tropical Americas and West Indies. The fruit is round to conical, 5–10 cm (2.0–3.9 in) in diametre and 6–10 cm (2.4–3.9 in) long, and weighing 100–240 g (3.5–8.5 oz), with a thick rind composed of knobby segments. The colour is typically pale green to blue-green, with a deep pink blush in certain varieties, and typically has a bloom. It is unique among *Annona* fruits in being segmented, and the segments tend to separate when ripe, exposing the interior.

The flesh is fragrant and sweet, creamy white to light yellow, and resembles and tastes like custard. It is found adhering to 13–16 mm (0.51–0.63 in) long seeds to form individual segments arranged in a single layer around the conical core. It is soft, slightly grainy, and slippery. The hard, shiny seeds may number 20-38 or more per fruit, and have a brown to black coat, although varieties exist that are almost seedless.

There are also new varieties being developed in Taiwan. The atemoya or "pineapple sugar-apple", a hybrid between the Sugar Apple and the Cherimoya, is popular in Taiwan, although it was first developed in the USA in 1908. The fruit is similar in sweetness to the sugar apple but has a very different taste. Like the name suggests, it tastes like pineapple. The arrangement of seeds is in spaced rows, with the fruit's flesh filling most of the fruit and making grooves for the seeds, instead of the flesh only occurring around the seeds.

Nomenclature

Due to its widespread cultivation, many local names have developed for the fruit.

In English, it is most widely known as a sugar apple or sweetsop as well as a custard apple, especially in India and Australia (custard apple also refers to *Annona reticulata*, a closely related species).

Sugar-apple is high in calories, an excellent source of vitamin C and manganese, a good source of thiamine and vitamin B_6, and provides vitamin B_2, B_3 B_5, B_9, iron, magnesium, phosphorus and potassium in fair quantities.

Production Practices

The atemoya tree is relatively small (10 m), with an open canopy. Leaves are elliptic, ovate, or lanceolate, and often variable in shape on the same tree. Leaf length is ten to 20 cm and width ranges from four to eight cm. Flowers are three to four cm in diametre with three fleshy, pale, yellowish-green petals. They are borne singly or in clusters of two to three in axils of leaves on year-old wood or on new shoots. The main bloom period in Florida is May through June, with a minor bloom in August through September. The flowers are protogyrous, functioning first as female, then as male. Atemoya bears an aggregate fruit from five to six months after bloom that may weigh up to one pound. It is at first green, but becomes more light green or yellowish green upon maturity. The pulp is white with a custard-like consistency and a sweet, pleasant flavour. There are typically ten to 40 seeds per fruit.

The sugar apple tree is also small (5 m), with a rounded canopy and slender branches. Thin, lanceolate leaves range in length from six to ten cm. The leaves are a dull, pale green with sparse hairs at flush, but become smooth at maturity. Flowers are about three cm in length and are produced singly or in clusters of two to four in late spring. The aggregate fruit produced is round, conical, or heart-shaped. The skin is thick and yellowish green. The pulp is creamy white and filled with numerous small, dark seeds.

Most sugar apple trees grown in Florida are seedlings since there is little variability among them and they grow true to type. When grafting is done, veneer grafting on custard apple (*Anonna reticulata*) is the preferred method. Shield budding also produces acceptable trees. A number of selections have been made from atemoya seedling populations. 'Gefner' produces satisfactory fruit without hand pollination. Other varieties either split upon maturity or produce few fruit without hand pollination. Atemoya grown from seed are extremely variable. Superior trees can be propagated by budding or grafting sugar apple, custard apple, or pond apple (*Annona glabra*). Chip budding or veneer grafting are the two most successful methods.

Both atemoya and sugar apple are intolerant of salty or freezing conditions, although mature trees may be able to withstand a few hours at sub-freezing temperatures. The trees become leafless during periods of cold. Both plants are adapted to moderate, evenly distributed rainfall. Leaf shedding is a mechanism used by both species to withstand periods of drought. The trees grow and bear best on neutral, fertile soils. Trees should be spaced five to six m apart between and within rows. Pruning for sugar apple is needed to remove dead wood and broken branches. However, atemoya is pruned just before the growing period to promote an open-centred form with few main branches.

Atemoya is self-fertile but individual flowers usually are not self-pollinated because stigmas are no longer receptive when the pollen is shed. Poor fruit set is a serious problem in some areas, evidently because of poor pollination Incomplete pollination will also result in misshapen fruit. Those flowers that open during warm, humid weather are more likely to set fruit than those borne when conditions are dry and/or cold. Hand pollinating when the female flower first opens increases fruit set in these problem areas

Insect/Mite Management

Insect/Mite Pests

The principal pest on atemoya and sugar apple in Florida is annona seed borer, while other pests such as ambrosia beetles, scales, mealybugs, aphids, and lepidopteran larvae are minor.

Annona Seed Borer (Bephratelloides cubensis)

This chalcidoid wasp was introduced into Florida in the 1920s. The female wasps lay eggs singly in small, immature fruit. The larvae feed on seed endosperm and pupate within the seed. Adult wasps emerge by burrowing to the fruit's exterior, which often leads to fungal infection and mummification of the fruit. The generation time for the wasp is about three months, and the adults live no more than two weeks.

Both atemoya and sugar apple are attacked by this wasp. However, since atemoya blooms and fruits slightly earlier in the spring than sugar apple, it is generally infected at a higher rate than is sugar apple. Peak infestation rates for atemoya and sugar apple in southern Florida in late summer were approximately 80 and 20 percent, respectively. The wasp overwinters in bullock's heart (*Annona reticulata*) and then attacks early-season fruit, while late-season fruit is attacked by newly emergent wasps.

Chemical Control

Twenty-five percent of atemoya and sugar apple growers who responded to a survey reported that they used insecticide. Those survey respondents who provided insect damage estimates indicated that from 5 to 100 percent of the crop would be lost to insect damage (n=5, mean of 62 percent). Insecticides and miticides registered for use on Florida atemoya and sugar apple in 2008 include azadirachtin, *Bacillus thuringiensis, Beauveria bassiana*, buprofezin, imidacloprid, insecticidal oil, insecticidal soap, methidathion, pyrethrin +/- rotenone, pyriproxyfen, spinetoram, spinosad, and sulfur. Pyriproxyfen and methoprene are available for fire ant control.

In non-bearing atemoya and sugar apple, bifenazate and bifenthrin are registered while pymetrozine, fenpropathrin, and hexythiazox are labelled for non-bearing nursery trees. Hydramethylnon and fenoxycarb are available for fire ant control in non-bearing atemoya and sugar apple.

Crop Oils

Crop oils work by smothering poorly mobile insects such as scales, aphids, and mites. The oils are usually made up as 1.5 to 3 percent solutions, which are applied thoroughly to each tree. The price of crop oils varies based on the amount and the formulation used, but an average application cost is $6/acre. The restricted entry interval (REI) for crop oils is 4 hours, and there is no pre-harvest interval. Thirteen percent of surveyed atemoya and sugar apple growers in Florida applied crop oil twice per season.

Azadirachtin

Azadirachtin is a natural compound derived from the neem tree (*Azadirachta indica*) that acts as an insect growth regulator and a deterrent. The compound is used to manage lacewing bugs and mealy bugs. The price of azadirachtin is $2100 per pound of active ingredient and the approximate cost per application is $90 per acre. The REI is 4 hours. Thirteen percent of atemoya and sugar apple growers applied azadirachtin to their acreage twice a season.

Cultural Control

Survey results of all respondents who grew tropical fruit show that 44 percent reported keeping records of pest problems, 50 percent adjusted applications (timing or rate) to protect beneficial insects and mites, and 52 percent alternated pesticides to reduce resistance. Sixty-two percent reported selecting the pesticide that is least toxic to beneficial insects and mites, and 63 percent spot sprayed only infested plants or areas. Seventy percent reported selecting pesticides that are least toxic to the environment to make this the dominant form of cultural pest control.

Biological Control

Seven percent of the responding tropical fruit growers reported release of predatory wasps for control of lepidopteran pests. Additionally, 30 percent reported the use of biologically derived pesticides like *B.t.*

Weed Management

Weed Pests

Weeds can reduce yields by competing primarily for water and nutrients. Although individual weed species may vary regionally, predominant weed species in groves are grasses, sedges, and pigweeds.

Chemical Control

There are few herbicides labelled for use on bearing atemoya and sugar apple (glyphosate, oxyfluorfen, carfentrazone, and pelargonic acid). Flumioxazin can be used on non-bearing trees. Eighty-eight percent of growers surveyed reported herbicide use.

Glyphosate

Glyphosate is a phosphorylated amino acid herbicide used for total vegetation control. Glyphosate is applied as a directed spray so that foliage is not injured. The median price of glyphosate is $10 per pound of active ingredient, and the approximate cost per application is $20 per acre for annual weeds and $50 per acre for perennial weeds. The REI for glyphosate is 4 hours and the PHI is 14 days. Eighty-eight percent of surveyed growers in Florida applied glyphosate either three (14 percent), four (43 percent), five (14 percent), or six (29 percent) times, for an average use of 4.6 times.

Disease Management

Disease Pathogens: The principal diseases affecting atemoya and sugar apple production in Florida include anthracnose (*Colletotrichum gloeosporioides*) and rust (*Phakopsora cherimoliae*). Fungal diseases that are non-manageable and require removal and destruction include mushroom root rot (*Armillaria tabescens*) and pythium root rot.

Anthracnose (caused by Colletotrichum gloeosporioides)

This fungus infects flowers, leaves, and fruit. Infected flowers develop dark lesions on the petals that enlarge and finally blacken and then kill the whole flower. Infected young fruit rot and mummify on the tree, but with mature fruit, the damage is mostly cosmetic: a small, shallow area of hardened tissue. Infected leaves develop light green lesions that enlarge into irregular brown areas, giving the leaves a scorched appearance. It can also cause premature leaf drop.

Rust (caused by Phakopsora cherimoliae)

This fungus infects leaves during the summer through fall. Slightly yellow spots develop in response to the infection. The fungus reproduces rapidly in the form of a slightly raised pustule or blister. Severe levels of infection can cause defoliation.

Chemical Control

Sixty-three percent of surveyed atemoya and sugar apple growers reported fungicide use. Those survey respondents that provided damage estimates indicated that from 10 to 100 percent of the crop would be lost to disease (n=4, mean of 55 percent). Fungicides registered for use on atemoya and sugar apple include azoxystrobin, copper hydroxide/ sulfate, mefenoxam, hydrogen dioxide, *Bacillus subtilis*, phosphoric acid, sulfur, and *Trichoderma harzianum*.

Copper

Copper has long been used as a fungicide, primarily to manage anthracnose. It can be applied in multiple forms (copper hydroxide, copper sulfate, etc.). The median price of copper hydroxide is $2 per pound of active ingredient. For atemoya, the approximate cost per application is $5 per acre. The cost per acre for sugar apple is $20. The REI for copper hydroxide/copper sulfate is 24 hours. Sixty-three percent of surveyed atemoya and sugar apple growers in Florida applied copper. These growers applied copper once (40 percent), twice (20 percent), three times (20 percent), or four times (20 percent) for an average use of 2.2 times per season.

Nematode Management

Nematode Pests: Plant-parasitic nematodes are microscopic roundworms found in the soil. They tend to attack plant roots. General signs of nematode damage include stunting, premature wilting, leaf yellowing, root malformation, and related signs characteristic of nutrient deficiencies. Stunting and poor stand development tend to occur in patches throughout the grove as a result of the irregular distribution of nematodes within the soil.

Chemical Control

There are no currently registered nematicides for use on atemoya or sugar apple. None of the surveyed growers reported the use of nematicides

Barbados Cherry

Malpighia emarginata is a tropical fruit-bearing shrub or small tree in the family Malpighiaceae. Common names include acerola (Portuguese pronunciation: Barbados cherry, West Indian cherry and wild crepe myrtle. Acerola is native to South America, southern Mexico, and Central America, but is now also being grown as far north as Texas and in subtropical areas of Asia, such as India. It is known for being extremely rich in vitamin C, almost as much as camu camu, although it also contains vitamins A, B_1, B_2, and B_3, as well as carotenoids and bioflavonoids, which provide important nutritive value and have antioxidant uses. The vitamin C produced by the fruit is better absorbed by humans than synthetic ascorbic acid.

Distribution

M. emarginata is originally from Yucatán, and can be found in Mexico, Central America, the Caribbean, South America as far south as Peru, and the southeast region of Brazil, and in the southernmost parts of the contiguous United States (southern Florida and the Lower Rio Grande Valley of Texas). In Florida, it can be grown in protected locations as far north as Cape Canaveral. It is cultivated in the tropics and subtropics throughout the world, including the Canary Islands, Ghana, Ethiopia, Madagascar, Zanzibar, Sri Lanka, Taiwan, India, Java, Hawaii, and Australia.

Adaptation

Acerola can be propagated by seed, cutting, or other methods. It prefers dry, sandy soil and full sun, and cannot endure temperatures lower than 30°F. Because of its shallow roots, it has very low tolerance to winds.

Description

Acerola is an evergreen shrub or small tree with spreading branches on a short trunk. It is usually 2–3 m (6.6–9.8 ft) tall, but sometimes reaches 6 m (20 ft) in height.

Figure: *Close-up of the blossom and unripe fruit*

Leaves

The leaves are simple ovate-lanceolate, 2–8 cm (0.79–3.15 in) long, 1–4 cm (0.39–1.57 in), and are attached to short petioles. They are opposite, ovate to elliptic-lanceolate, and have entire or undulating margins with small hairs, which can irritate skin.

Flowers

Flowers are bisexual and 1–2 cm (0.39–0.79 in) in diametre. They have five pale to deep pink or red fringed petals, 10 stamens, and six to 10 glands on the calyx. The three to five flowers per inflorescence are sessile or short-peduncled axillary cymes.

Fruit

After three years, trees produce significant numbers of bright red drupes 1–3 cm (0.39–1.18 in) in diametre with a mass of 3–5 g (0.11–0.18 oz). Drupes are in pairs or groups of three, and each contains three triangular seeds. The drupes are juicy and very high in vitamin C (3-46 g/kg) and other nutrients. They are divided into three obscure lobes and are usually acidic to subacidic, giving them a sour taste, but may be sweet if grown well. While the nutrient composition depends on the strain and environmental conditions, the most common components of acerola and their concentration ranges, per 100 g, are: proteins (2.1-8.0 g), lipids (2.3-8.0 g), carbohydrates (35.7-78 g), calcium (117 mg), phosphorus (171 mg), iron (2.4 mg), pyridoxine (87 mg), riboflavin (0.7 mg), thiamine (0.2 mg), water (906-920 g) and dietary fibre (30 g)

Uses

As food: The fruit is edible and widely consumed in the species' native area, and is cultivated elsewhere for its high vitamin C content. There are 1677.6 mg of vitamin C in 100 g of fruit. The fruit can be used to make juices and pulps, vitamin C concentrate, and baby food, among other things.

A comparative analysis of antioxidant potency among a variety of frozen juice pulps was carried out, including the acerola fruit. Among the 11 fruit pulps tested, acerola was the highest-scoring domestic fruit, meaning it had the most antioxidant potency, with a Trolox equivalent antioxidant capacity score of 53.2 mg. Absolut Vodka released Absolut Los Angeles, a limited-edition spirit flavored with acerola, açai, pomegranate, and blueberry, in July 2008.

Other uses

Acerola is a popular bonsai subject because of its small leaf, fruit, and fine ramification. It is also grown as an ornamental and for hedges. It is one of three ingredients in a proprietary herbal medicine for allergic rhinitis.

Ecology

M. emarginata is a host plant for the caterpillars of the white-patched skipper (*Chiomara asychis*), Florida duskywing (*Ephyriades brunneus*), and brown-banded skipper (*Timochares ruptifasciatus*). Larvae of the acerola weevil (*Anthonomus macromalus*) feed on the fruits, while adults consume young leaves.

Caimito (Star Apple)

Chrysophyllum cainito is a tropical tree of the family Sapotaceae. It is native to the Greater Antilles and the West Indies. It has spread to the lowlands of Central America and is now is grown throughout the tropics, including Southeast Asia. It grows rapidly and reaches 20 m in height.

Overview

It has numerous common names including cainito, caimito, star apple, golden leaf tree, abiaba, pomme du lait, estrella, milk fruit and aguay. It is also known by the synonym *Achras cainito*. In Vietnam, it is called vú sïa (literally: breast-milk). In Malayalam it is called "Swarnapathry" means '[the tree with] golden leaves'.

The leaves are evergreen, alternate, simple oval, entire, 5–15 cm long; the underside shines with a golden colour when seen from a distance. The tiny flowers are purplish white and have a sweet fragrant smell. The tree is also hermaphroditic (self-fertile). It produces a strong odor.

Fruits

Figure: *Fruits are also available in green or red*

The fruit is globose and typically measures from 2 to 3 inches in diametre. When ripe, it usually purple-skinned with a faint green area appearing around the calyx. A radiating star pattern is visible in the pulp. Greenish-white and yellow-fruited cultivars are sometimes available. The skin is rich in latex, and both it and the rind are not edible. The flattened seeds are light brown and hard. It is a seasonal fruit bearing tree.

The fruits are delicious as a fresh dessert fruit; it is sweet and best served chilled. Infusions of the leaves have been used against diabetes and articular rheumatism. The fruit has anti-oxidant properties., The bark is considered a tonic and stimulant, and a bark decoction is used as an antitussive. The fruit also exists in three colours, dark purple, greenish brown and yellow. The purple fruit has a denser skin and texture while the greenish brown fruit has a thin skin and a more liquid pulp; the yellow variety is less common and difficult to find.

A number of closely related species, also called star apples, are grown in Africa, including *C. albidum* and *C. africanum*.

In Sierra Leone the fruit is referred to as "Bobi wata" or breast milk fruit.

Origin and Distribution

It is commonly stated that the star apple is indigenous to Central America but the eminent botanists Paul Standley and Louis Williams have declared that it is not native to that area, no Nahuatl name has been found, and the tree may properly belong to the West Indies. However, it is more or less naturalized at low and medium altitudes from southern Mexico to Panama, is especially abundant on the Pacific side of Guatemala, and frequently cultivated as far south as northern Argentina and Peru. It was recorded by Ciezo de Leon as growing in Peru during his travels between 1532 and 1550. It is common throughout most of the Caribbean Islands and in Bermuda. In Haiti, the star apple was the favourite fruit of King Christophe and he held court under the shade of a very large specimen at Milot. The United States Department of Agriculture received seeds from Jamaica in 1904 (S.P.I. #17093). The star apple is grown occasionally in southern Florida and in Hawaii where it was introduced before 1901. There are some trees in Samoa and in Malaya though they do not bear regularly. The tree is grown in southern Vietnam and in Kampuchea for its fruits but more for its ornamental value in West Tropical Africa, Zanzibar, and the warmer parts of India. It was introduced into Ceylon in 1802, reached the Philippines much later but has become very common there as a roadside tree and the fruit is appreciated.

Varieties

Apart from the two distinct colour types, there is little evidence of such pronounced variation that growers would be stimulated to make vigorous efforts to select and propagate superior clones. William Whitman of Miami observed a tree yielding heavy crops of well-formed, high quality fruits in Port-au-Prince, Haiti, from late January to the end of

June. He brought budwood to Florida in 1953. Grafted progeny and trees grown from air-layers have borne well here even prior to reaching 10 ft (3 m) in height. This introduction, named the "Haitian Star Apple", is propagated commercially for dooryard culture. Seeds of the Port-au-Prince tree have produced seedlings that have performed poorly in Florida.

Climate

The star apple tree is a tropical or near-tropical species ranging only up to 1,400 ft (425 m) elevation in Jamaica. It does well only in the warmest locations of southern Florida and on the Florida Keys. Mature trees are seriously injured by temperatures below 28° F (-2.22° C) and recover slowly. Young trees may be killed by even short exposure to 31° F (-0.56° C).

Soil

The tree is not particular as to soil, growing well in deep, rich earth, clayey loam, sand, or limestone, but it needs perfect drainage.

Propagation

Star apple trees are most widely grown from seeds which retain viability for several months and germinate readily. The seedlings bear in 5 to 10 years. Vegetative propagation hastens production and should be more commonly practise. Cuttings of mature wood root well. Air-layers can be produced in 4 to 7 months and bear early. Budded or grafted trees have been known to fruit one year after being set in the ground. In India, the star apple is sometimes inarched on star apple seedlings. Grafting on the related satinleaf tree *(C. oliviforme* L.) has had the effect of slowing and stunting the growth.

Culture

During the first 6 months, the young trees should be watered weekly. Later irrigation may be infrequent except during the flowering season when watering will increase fruit-set. Most star apple trees in tropical America and the West Indies are never fertilized but a complete, well-balanced fertilizer will greatly improve performance in limestone and other infertile soils.

Harvesting

Star apples are generally in season from late winter or early spring to early summer. They do not fall when ripe but must be hand-picked by clipping the stem. Care must be taken to make sure that they are fully mature. Otherwise the fruits will be gummy, astringent and inedible. When fully ripe, the skin is dull, a trifle wrinkled, and the fruit is slightly soft to the touch.

Yield

In India, a mature star apple tree may bear 150 lbs (60 kg) of fruits in the short fruiting season of February and March.

Keeping Quality: Ripe fruits remain in good condition for 3 weeks at 37.4° to 42.8° F (3°-6° C) and 90% relative humidity.

Pests and Diseases

Larvae of small insects are sometimes found in the ripe fruits.

The, main disease problem in the Philippines is stem-end decay caused by species of *Pestalotia* and *Diplodia.* In Florida, some fruits may mummify before they are full-grown.

The foliage is subject to leaf spots from attack by *Phomopsis* sp., *Phyllosticta* sp., and *Cephaleuros virescens,* the latter known as algal leaf spot or green scurf.

Birds and squirrels attack the fruits if they are left to fully ripen on the tree.

Longan

Dimocarpus longan, commonly known as the longan, is a tropical tree that produces edible fruit. It is one of the better-known tropical members of the soapberry family, to which the lychee also belongs. It is native to Southern Asia.

The longan (pinyin: *lóng yÎn*; literally: "Dragon Eye"), is so named because it resembles an eyeball when its fruit is shelled (the black seed shows through the translucent flesh like a pupil/iris). The seed is small, round and hard, and of an enamel-like, lacquered black. The fully ripened, freshly harvested shell is bark-like, thin, and firm, making the fruit easy to shell by squeezing the fruit out as if one is "cracking" a sunflower seed. When the shell has more moisture content and is more tender, the fruit becomes less convenient to shell. The tenderness of the shell varies due to either premature harvest, variety, weather conditions, or transport/storage conditions. In China, it is also called *guì*, especially when dried.

Tree Description

The *Dimocarpus longan* tree is a medium-sized evergreen that can grow up to 6 to 7 metres (20 to 23 ft) in height. It is somewhat sensitive to frost. Longan trees prefer sandy soil. While the species prefers temperatures that do not typically fall below 4.5 °C (40 °F), it can withstand brief temperature drops to about –2 °C (28 °F). Longans and lychees bear fruit at around the same time of the year.

The longan was listed as Near Threatened on the IUCN Red List.

Culinary Uses

The fruit is sweet, juicy and succulent in superior agricultural varieties and, apart from being eaten fresh, is also often used in East Asian soups, snacks, desserts, and sweet-and-sour foods, either fresh or dried, sometimes canned with syrup. The taste is different from lychees; while longan have a drier sweetness, lychees are often messily juicy with a more tropical, sour sweetness.

The seed and the shell are not consumed.

Dried longan are often used in Chinese cuisine and Chinese sweet dessert soups. In Chinese food therapy and herbal medicine, it is believed to have an effect on relaxation. In contrast with the fresh fruit, which is juicy and white, the flesh of dried longans is dark brown to almost black. In Chinese medicine, the longan, much like the lychee, is thought to give internal "heat" (Nkp).

Longans, Raw

Cultivation: Potassium chlorate has been found to cause the longan tree to blossom. However, this causes stress on the tree if it is used excessively, and eventually kills it. It seems that the type of longan originally brought to the New World was not one of the best, having aroused so little interest in the fruit. Groff stated that the leading variety of Fukien was the round-fruited 'Shih hsía', the "Stone Gorge Lungan" from P'ing Chou. There were 2 types, one, 'Hei ho shih hsia', black-seeded, and 'Chin ch' i ho shih hsia', brown-seeded. This variety did not excel in size but the flesh was crisp, sweeter than in other varieties, the seed small and the dried flesh, after soaking in water, was restored almost to fresh condition.

None of the other 4 varieties described by Groff has any great merit.

'Wu Yuan' ("black ball") has small, sour fruit used for canning. The tree is vigorous and seedlings are valued as rootstocks. 'Kao Yuan' is believed to be a slightly better type of this variety and is widely canned.

'Tsao ho' ('Early Rice') is the earliest variety and a form called 'Ch'i chin tsao ho' precedes it by 2 weeks. In quality, both are inferior to 'Wu Yuan'.

'She p' i' ('Snake skin') has the largest fruit, as big as a small lychee and slightly elongated. The skin is rough, the seed large, some of the juice is between the rind and the flesh, and the quality is low. Its only advantage is that it is very late in season.

'Hua Kioh' ('Flower Skin'), slightly elongated, has thin, nearly tasteless flesh, some of the juice is between the rind and the flesh, and the overall quality is poor. It is seldom propagated vegetatively.

There are no "chicken- tongue" (aborted seed) varieties in China.

There are 2 improved cultivars grown extensively in Taiwan– 'Fukien Lungan' ('Fukugan') was introduced from Fukien Province in mainland China. The other, very similar and possibly a mutant of 'Fukien', is 'Lungan Late', which matures a month later than 'Fukien'.

In 1954, William Whitman of Miami introduced a superior variety of longan, the 'Kohala', from Hawaii. It began to bear in 1958. The fruit is large for the species, the seed is small, and the flesh is aromatic, sweet and spicy. The tree produces fairly good crops in midsummer. One hundred or more air-layers have been brought by air from Hawaii and planted at various locations in southern Florida and in the Bahamas. A seedling planting and selection programme was started in 1962 at the USDA Subtropical Horticulture Research Unit, Miami.

The plants were all open-pollinated seedlings of the canning variety, 'Wu Yuan', brought in from Canton in 1930 as P.I. #89409. Some set fruit in 1966 and 1967 but more of them in 1968. Evaluation of these and other acquisitions continues. Included in the study are M-17886, 'Chom Poo Nuch', and M-17887, 'E-Haw'.

Climate

Professor Groff wrote that "the lungan . . . is found growing at higher latitudes and higher altitudes than the lychee." Also: "On the higher elevations of the mountainous regions which are subject to frost the lychee is seldom grown. The longan appears in these regions more often but it, too, cannot stand heavy frosts." The longan's range in Florida extends north to Tampa on the west coast and to Merritt Island on the east coast. Still, small trees suffer leaf-and twig-damage if the temperature falls to 31° or 30° F (-0.56°—1.11° C) and are killed at just a few degrees lower. Larger trees show leaf injury at 27° to 28° F (-2.78°—2.22° C), small branch injury at 25° to 26° F (-3-89°—3.33° C), large branch and trunk symptoms at 24° F (-4.44° C) and sometimes fail to recover.

On the other hand, after a long period of cool weather over the 3 winter months, with no frost, longan trees bloom well. Blooming is poor after a warm winter.

Soil

The longan thrives best on a rich sandy loam and nearly as well on moderately acid, somewhat organic, sand. It also grows to a large size and bears heavily in oolitic limestone. In organic muck soils, blooming and fruiting are deficient.

Propagation

Most longan trees have been grown from seed. The seeds lose viability quickly. After drying in the shade for 4 day, they should be planted without delay, but no more than 3/4 in (2 cm) deep, otherwise they may send up more than one sprout. Germination takes place within a week or 10 days. The seedlings are transplanted to shaded nursery rows the following spring and set in the field 2-3 years later during winter dormancy.

In Kwangtung Province, when vegetative propagation is undertaken, it is mostly by means of inarching, nearly always onto 'Wu Yuan' trees 3-5 years old and 5 to 6 ft (1.5-1.8 m) high. The union is made no less than 4 ft (1.2 m) from the ground because it is most convenient. Nevertheless, the point of attachment remains weak and needs to be braced with bamboo to avoid breaking in high winds.

Grafting is uncommon and when it is done, it is a sandwich graft on longan rootstock, 3 or 4 grafts being made successively, one onto the beheaded top of the preceding one, in the belief that it makes the graft wind-resistant and that it induces better size and quality in the fruit. Conventional modes of grafting have not been successful in Florida, but whip-grafting has given 80% success in Taiwan. Air-layering is frequently done in Fukien Province and was found a feasible means of distributing the 'Kohala' from Hawaii. Air-layers bear in 2 to 3 years after planting. A tree can be converted to a preferred cultivar by cutting it drastically back and veneer-grafting the new shoots.

Culture

In China, if the longan is raised on the lowlands it is always put on the edges of raised beds. On high ground, the trees are placed in pre-enriched holes on the surface. The trees are fertilized after the fruit harvest and during the blooming season, at which time the proportion of nitrogen is reduced. Fresh, rich soil is added around the base of the trees year after year. The longan needs an adequate supply of water and can even stand brief flooding, but not prolonged drought. Irrigation is necessary in dry periods.

An important operation is the pruning of many flower-bearing twigs–3/4 of the flower spikes in the cluster being removed. Later, the fruit clusters are also thinned, in order to increase the size and quality of the fruits.

Generally, the trees are planted too close together, seriously inhibiting productivity when they become overcrowded. In China, full-grown trees given sufficient room–at least 40 ft (12 m) apart–may yield 400 to 500 lbs (180-225 kg) in good years. Crops in Florida from trees 20 ft (6 m) tall and broad, have varied from light–50-100 lbs (22.5-45 kg)–to medium–150-250 lbs (68-113 kg), and heavy–300-500 lbs (135-225 kg). Rarely such trees may produce 600-700 lbs (272-317 kg). Larger trees have larger crops but if the trees become too tall harvesting is too difficult, and they should be topped. Harvesters, working manually from ladders, or using pruning poles cut the entire cluster of fruit with leaves attached.

A serious problem with the longan is its irregular bearing–often one good year followed by 1 or 2 poor years. Another handicap is the ripening season–early to mid-August in China, which is the time of typhoons; August and September in Florida which is during the hurricane season. Rain is a major nuisance in harvesting and in conveying the fruit to market or to drying sheds or processing plants.

Keeping Quality

At room temperature, longans remain in good condition for several days. Because of the firmer rind, the fruit is less perishable than the lychee. Preliminary tests in Florida indicate that the fruit can be frozen and will not break down as quickly as the lychee when thawed.

Pests and Diseases

The longan is relatively free of pests and diseases. At times, there may be signs of mineral deficiency which can be readily corrected by supplying minor elements in the fertilization programme.

Major Pests

Regardless of where lychee is grown, several insect groups attack the flowers, fruit, leaves and branches. Lepidopterous fruit borers are generally the most important pests affecting production. Other important species include various leaf- and flower-eating caterpillars and beetles, bark borers, scales, leaf mites, fruit-sucking bugs, fruit-piercing moths and fruit flies.

Fruit borers

Conopomorpha sinensis Bradley, known as the lychee stem-end borer in China and the lychee fruit borer in Thailand, is the major pest in most seasons. This pest was previously recorded as *Acrocercops cramerella* (now *Conopomorpha cramerella* Snellen). *C. sinensis* and the related *C. litchiella* Bradley both attack lychee, the latter preferring leaves and shoots, while *C. cramerella* is restricted to rambutan and cocoa (Bradley, 1986).

C. sinensis lays yellow, scale-like eggs 0.4 x 0.2 mm long on the fruit any time after flowering, as well as on new leaves and shoots. Both lychee and longan are affected. The eggs hatch in three to five days, with the larva immediately penetrating the fruit, leaf or shoot. They tunnel through the flesh of the fruit that often fall from the tree.

In Thailand, fruit are inspected weekly from fruit set to detect eggs of *C. sinensis*, which are very small and almost invisible to the naked eye. Infested fruit should be picked and destroyed, at infestation levels of 1 to 2 percent. When the pest becomes more active, permethrin is applied weekly, up to two weeks before harvest. In Taiwan Province of China, cypermethrin, deltamethrin, carbofuran or fenthion during early fruit set is recommended to prevent damage later in the season. Moths can be excluded by enclosing the fruit panicles in nylon mesh bags, but is uneconomic in areas with high labour costs. If the parasitoids *Phanerotoma* sp. and *Apanteles* sp. are not active, fallen fruit should be removed to reduce the build-up of moths.

All stages of the leaf-miner, *Conopomorpha litchiella* Bradley, are similar to those of the fruit borer. The female lays its eggs on new shoots and the small, light-yellow eggs hatch three to five days later. The newly hatched larva is creamy white, and bores into shoots and leaf blades, usually into the mid-rib and veins. The moths are attracted to leaf flushes that emerge during the rainy season from June to October in Thailand. Affected shoots often wilt.

Bearing trees should be inspected during early flush development and sprayed if necessary. The leaf flush before flower initiation is very important as it supplies the carbohydrates needed for fruit development. If 30 to 40 percent of the larvae are parasitised, spraying is not recommended. Young, non-bearing trees do not need to be sprayed either. This also allows the parasitoids to build up in the orchard.

The insect originally referred to as *Argyroploce illepida* Butler (= *Cryptophlebia carpophaga* Walsingham) in India (Butani, 1977), is actually *Cryptophlebia ombrodelta* Lower (Bradley, 1953). It also occurs in Thailand, China, Japan, Taiwan Province of China and Australia, but only in the latter area is it regarded as a significant pest.

The creamy white eggs of these species are oval and flat with a reticulate surface, and are about 1.0 x 0.8 mm. They are laid singly or in groups of up to 15 on the fruit surface. The newly hatched larva feeds on the fruit skin and then tunnels towards the seed. In immature fruit, the young larva bores directly into the seed, which is completely eaten. A single larva may damage two or three fruit, if the fruit are small. However, they prefer mature colouring fruit with larger seeds.

In South Africa, the insect growth regulator, triflumuron as a single, full cover spray 40 days before harvest, or two sprays of teflubenzuron a fortnight apart commencing when the fruit are 10 mm in diametre, are recommended. Alternatively, the panicles can be covered with paper bags. The bags also improve fruit colour and overall quality. In Queensland, carbaryl and azinphos-methyl have been used with varying success. Several sprays commencing at fruit colouring are applied on a calendar basis, with monitoring for the presence of eggs less common. Newer insecticides including the insect growth regulator, tebufenozide, provide better control, with less disruption to natural enemies.

The various species of *Cryptophlebia* are attacked by their own complex of egg, larval and pupal parasitoids; however, these do not always keep borers below economic thresholds. Egg parasitoids such as *Trichogrammatoidea fulva* Nagaraja from India and *T. cryptophlebiae* from South Africa and Australia, offer the best prospects for biological control.

Fruit-piercing Moths

Fruit-piercing moths such as *Eudocima (Othreis) fullonia* (Clerck), *Eudocima salaminia* (Cramer) and *Eudocima jordani* (Holland) are important throughout Asia, Australia and the South Pacific. The larvae of these moths develop on a variety of host plants such as the coral tree, *Erythrina*, and vines of the Menispermaceae (*Legnephora, Stephania, Fawcettia, Tinospora, Carronia, Sarcopetalum, Pleogyne* and *Hypserpa*).

The moths have a proboscis that drills a neat hole in the skin of the fruit allowing them to suck the juice from the flesh. Contamination of the wound with yeasts and bacteria carried on the proboscis destroys the fruit. *Drosophila* spp. attracted to the fermenting juice hastens deterioration. Within a few days, a frothy exudate seeps from the fruit and stains undamaged fruit close by. In Australia and Thailand, farmers go to their orchards at night with spotlights and attempt to catch the moths. However, this is futile.

Australian farmers also make traps by draping shade cloth loosely over a frame of wire and baiting it with fermenting citrus and bananas.

The moths feed on the fruit and become entangled in the folds of the shade cloth when they attempt to fly off. Several traps are required to protect an orchard and even then, substantial damage is sustained. In Thailand, ripe bananas and pineapples are dipped in insecticide and hung in the trees to poison the feeding moths. In some countries, panicles are covered with paper bags. In recent times, parrots and fruit bats have become a severe problem for growers in Australia. Fine nets erected over the orchard control fruit-piercing moths as well as the vertebrate pests.

Leaf-feeding Caterpillars

Oxyodes scrobiculata F. and *Oxyodes tricolour* Guen. occupy similar niches in Thailand and Australia. In Australia, *O. tricolour* attacks trees in southern Queensland, but is not a pest in the north. The castor oil looper, *Achaea janata* (L.), is a voracious feeder in Australia and often infests trees in large numbers at the same time as *O. tricolour*. The caterpillars can cause severe defoliation.

In Thailand it is recommended that carbaryl be applied when there are two to three young larvae per leaflet. Shaking the tree to dislodge larvae onto the ground improves the effectiveness of the sprays. If 40 percent or more of larvae are parasitised, sprays should not be necessary. In Australia, *Bacillus thuringiensis* Berliner (Bt), endosulfan or methomyl may be used when damaging populations of *O. tricolour* appear.

Leafrollers

Olethreutes perdulata Meyr. is an occasional pest in Queensland. *Platypeplus aprobola* (Meyrick) has also been recorded in Australia, China and India. *Adoxophyes cyrtosema* Meyr. and *Homona coffearia* Nietner attack trees in Guangzhou and Fujian. The latter species along with *Homona difficilis* is recorded in lychee, longan and rambutan in Thailand. The orange fruit borer, *Isotenes miserana* (Walker), is another leafroller that also attacks flowers and fruit in Queensland.

P. aprobola is a minor pest in China and India where it attacks leaves and flowers. However in Australia, it is part of a complex of species contributing to a significant loss of flowers. In China, *A. cyrtosema* and *H. coffearia* also feed on leaves, flowers and fruit.

In Australia, the damage caused by leafrollers is tolerated so long as it is restricted to the foliage and unlikely to affect flower initiation. If necessary, methomyl or carbaryl can be applied when 20 percent of leaf flushes are infested, to minimize damage to young trees or at critical periods of leaf growth in older trees. In India, rolled leaves that contain larvae are removed manually during light infestations, but phosphamidon, fenitrothion or endosulfan are applied for heavy infestations.

Beetle Borers

The longicorn beetle, *Aristobia testudo* (Voet), is a serious pest of lychee and longan in Guangdong (Zhang, 1997). The beetle has one generation per year, with adults emerging from June to August. The females girdle branches by chewing off 10 mm strips of bark, with the eggs laid on the wound and covered with exudate. The larvae hatch from late August and live under the bark until January when they bore into the xylem and create tunnels up 60 cm long.

In Taiwan Province of China, the white-spotted longicorn beetle, *Anoplophora maculata* (Thomson), has a one-year life cycle. Adults emerge in spring and females insert about twenty eggs individually into T-shaped incisions in the bark, 0.5 m above the soil surface. The larval period lasts about ten months. In Australia, the longicorn beetle, *Uracanthus cryptophagus*, causes similar damage (Plate 7).

Tunnelling by the larvae may kill branches, but rarely whole trees. Ring-barking of twigs by ovipositing adults causes the shoot tips to die and snap off. In China, regular inspections of trees during adult activity enable orchard workers to remove the beetles. Eggs and young larvae can also be removed at the same time. Established larvae can be located from the appearance of their frass, which is packed into the ends of tunnels. They can then be 'fished out' with wire hooks and knives. Alternatively, dichlorvos is injected and the tunnels sealed with clay (Zhang, 1997).

Scarab Beetles

The elephant beetle, *Xylotrupes gideon* (Linnaeus), is important in all areas of Australia. The larvae develop in the soil or mulch where they feed on plant roots and humus. The large, heavily-sclerotised and sexually-dimorphic adults emerge in spring. Later, they are attracted to the fruit as they ripen, especially those that have split or been damaged by parrots and fruit bats. They then start attacking sound fruit and can cause significant economic losses in the week or so leading up to harvest. Chemical control is unsatisfactory. Manual removal is effective in small trees, but difficult in large trees. Labour is relatively expensive in Australia, so this operation adds significantly to growing costs.

Soft Scales

Pulvinaria (*Chlorpulvinaria*) *psidii* (Maskell), the green shield scale, infests trees in China, Taiwan Province of China, Australia and

India. In Queensland, crawlers are produced in spring by adult scales that infest the leaves and twigs. Some of these crawlers move onto the flowers and young fruit. The female scales are sometimes mistaken for mealybugs because the egg masses that are covered in waxy filaments cover the ends of the scale.

Soft brown scale, *Coccus hesperidum* Linnaeus, is an occasional pest in Queensland, where chemicals have disrupted its parasitoids or it is protected by ants. *Parasaissetia nigra* (Nietner) and *Saissetia coffeae* (Walker) infest trees in India along with *C. psidii*, but they are not important. These scales cause no damage as they feed, but the fruit become unmarketable when significant populations develop on the surface, as they often do in China, Taiwan Province of China and Australia. The scales also produce honeydew, which supports the growth of sooty mould on infested fruit and panicles, and those below. These discoloured fruit are downgraded or rejected in the market-place.

Severe infestations may be controlled with methidathion, although applications of mineral oil are preferred so that effective predators, the mealybug ladybird, *Cryptolaemus montrouzieri* Mulsant, and the green lacewing, *Mallada signata* (Shneider) are not affected.

Bugs

Several bugs belonging to Tessaritomidae attack lychee and longan throughout Asia and Australia. *Tessaritoma papillosa* Drury occurs in southern China, Vietnam, Thailand, Myanmar, the Philippines and India, although Butani (1977) notes that *Tessaritoma javanica* Thunberg and *Tessaritoma quadrata* Distant, are the main species in India. In Australia, *Lyramorpha rosea* Westw. is known as the lychee stink bug, but rarely causes damage.

In China, *T. papillosa* has one generation per year. Adults tend to aggregate and over-winter mostly on lychee and longan, but may also be found on other hosts in warm protected areas. In spring, the females are attracted to trees with new flowers and shoots. They mate and lay up to 14 egg masses, each containing about 14 eggs, on the back of leaves. Peak egg-laying occurs in March in Guangdong, but continues through to September. The first nymphs mature in June, while there are still old adults in the trees. These old adults may have lived for up to a year, and generally die by August. The new adults do not mate immediately, but mature over winter and mate and lay eggs the following spring.

Adults and nymphs feed on terminals, which may be killed, and also on flowers and fruit, causing these to fall. Liu and Lai (1998) claimed that up to 30 percent of fruit in commercial orchards are damaged despite chemical applications.

In Guangdong, the main natural enemies are the egg parasitoids, *Encyrtus* (*Ooencyrtus*) sp., *Anastatus* sp. and *Blastophaga* sp. which parasitise 70 to 90 percent of eggs laid late in the season. Similar results were recorded by Liu and Lai (1998) when parasitised egg cards were hung in trees during March. In orchards under integrated pest management, combined parasitism rates by *Anastatus* sp. and *Ooencytus* sp. may reach 50 percent in June, but may be less than 3 percent in orchards that rely on chemicals. During the 1970s, biological control in Guangdong was initiated using the egg parasitoid *Anastatus japonicus* Ashmead, the flat venter wasp, after field trials had demonstrated its value. Since only 10 percent of eggs are parasitised by April when most of the eggs are laid, natural control is ineffective. In contrast, very good control with up to 90 percent parasitism is achieved after mass release of wasps.

In Thailand, the egg parasitoids *Anastatus* sp. nr. *japonicus* and *Ooencrytus phongi*, operate in a similar manner to their counterparts in China. Low levels of control are achieved during the critical early fruit production period, building up to good levels later. Mass rearing of the parasitoids in the wild silk worm, *Philosamia ricini* Hutt., and releasing them early, produced results similar to those in China. *Anastatus* sp. and *O. phongi* parasitised 79 percent and 21 percent of eggs, respectively (Nanta, 1992).

If chemicals are used, the timing of sprays is critical because the bugs vary in their susceptibility to trichlorfon at different times of the year, depending on body fat content and its nature.

Amblypelta nitida Stål, the fruitspotting bug, and *Amblypelta lutescens lutescens* (Distant), the banana spotting bug, are major pests of tropical fruit in Queensland (Waite, 1990). The adults over-winter on citrus or non-crop native or exotic ornamentals, and start to move into lychee and longan orchards in spring when the trees flower. They prefer to feed on green fruit, and so are very common just after fruit set. Orchards near rainforests where the bugs breed are particularly susceptible (Waite and Huwer, 1998).

The bugs feed on the developing seed and this causes the fruit to fall a couple of days later. The puncture mark is invisible on the fruit surface and the only way to distinguish damage from natural drop is to dissect the fruit. Fruit that has been attacked typically have a tan lesion on the seed testa. Endosulfan should be applied if more than 10 percent of fallen fruit have been stung. Usually, a maximum of two sprays applied two weeks apart, during the first six weeks after fruit set is sufficient.

Mites

Erinose mite, *Aceria litchii* (Keiffer), also known as hairy mite, hairy spider, or dog ear mite, occurs throughout China, Taiwan Province of China, India, Pakistan and Australia. Females lay eggs singly on the leaf surface amongst the erineum induced by their feeding. The eggs are only 0.032 mm in diametre, spherical and translucent white. The mites are also small, only 0.13mm long and pinkish-white. All stages have only four legs, but are quite mobile and move easily from old leaves to infest new flushes. Their feeding stimulates the production of the erineum where they shelter and feed. Numbers vary with the cycle of shoot growth, and are highest in summer and lowest in winter. Planting material obtained as marcots may be infested if they have been taken from trees with the mites. Later infestations occur when the mites are moved around the orchard by direct contact between trees, or carried around by orchard workers, wind and bees (Waite and McAlpine, 1992).

The mites attack new leaves causing a felt-like erineum to be produced on the under-surface. This forms as small blisters but may eventually covers the entire leaf, causing it to curl. In severe cases, whole terminals may be deformed. The young erineum is silver-white, changing to light brown and dark reddish-brown, and eventually black. The greatest numbers of mites are found in the intermediate stages.

Many leaves are ruined if infestations are severe. This generally causes no problems in established trees, but can debilitate young orchards. There can also be a problem if the mite moves from leaves onto the developing flowers and fruit. Fruit set can be disrupted or the fruit deformed. Such fruit are unmarketable. Numerous species of predatory mites, particularly those from the Phytoseidae, have been recorded with *A. litchii* (Wu *et al.*, 1991; Waite and Gerson, 1994). *Agistemus exsertus* Gonzalez (Stigmaeidae), has been used for control in Guangdong, Guangxi and Fujian.

Branches infested with the mite should be cut off and burnt. The mites can be controlled by applying insecticides when they move from the older leaves to a new flush. The leaves should be checked regularly for symptoms over summer and autumn. Not all trees in an orchard will be flushing or infested at the same time. In Australia, three sprays of dimethoate or wettable sulphur every two to three week during leaf emergence and expansion generally provide adequate control. Chemicals recommended in China include dichlorvos, dimethoate, dicofol, chlorpyrifos, omethoate and isocarbophos (Zhang, 1997).

Gall Flies

The leaf midge, *Dasyneura* sp., is a major pest in China (Zhang, 1997). *Litchiomyia chinensis* Yang and Luo was described from specimens reared from galls on lychee leaves collected in Guangdong. The larvae over-winter in the galls produced as a result of their feeding. They pupate in the soil, with the adult flies initiating the first of eight overlapping generations from March. The midges prefer damp, closed canopies and dry out in exposed areas. The adults lay eggs in lines on young leaves. The larvae then mine the leaf, causing 'watery dots' that later become the "galls". These turn brown and eventually drop out, giving the leaf a "shot-hole" appearance.

In susceptible orchards, monitoring is not effective and preventive procedures are required. As with erinose mite, infested leaves can be removed after harvest and burnt. Later in the spring, methyl parathion (2.5 percent) at 75 kg per ha can be distributed under the trees, or isofenphos (0.001 percent) can be sprayed on the ground just prior to emergence of the adults. In autumn, isocarbophos (0.001 percent) should be sprayed twice over two weeks during early leaf development (Zhang, 1997).

Fruit Flies

In Queensland, *Bactrocera tryoni* (Froggatt) occasionally attacks lychee, but is not considered economically significant. Females lay their eggs through the skin of the fruit, often utilizing cracks and wounds made by other pests. Although the eggs can hatch, the larvae rarely survive (de Villiers, 1992), probably because of the juice in mature fruit drowns them. The flies in Queensland, and related species in Africa and Hawaii, are capable of ovipositing through the skin of lychee, although some cultivars may be too thick. The only real fly problem appears to be in South Africa, with *Ceratatis rosa*. However, the level of damage is still quite low. In South Africa, pheromone-baited traps can be used for monitoring populations. Control is achieved with bait sprays of protein hydrolysate mixed with trichlorfon or mercaptothion. Alternatively, the panicles can be covered with paper bags after the November fruit drop.

Major Diseases

No major disease currently limits commercial production in Asia. Diseases are more important after harvest, although undoubtedly many of the fruit are infected before picking. There are a few organisms that infect the leaves, flowers and fruit, and a few others associated with tree decline and tree deaths. Chemicals are generally available for controlling diseases on the flowers and fruit. In contrast, more efforts need to be made to control the loss of trees.

Brown Blight

Brown blight, *Peronophythora litchii*, is a major disease in both lychee and longan in China and Thailand, although more important in the former. It is also reported to affect lychee in India. It is well described in Guangdong, and attacks leaves and panicles, as well as fruit that can be infected right up to harvest. These infections all reduce production and shelf life. Flower panicles are particularly susceptible. Immature fruit turn brown, while those infected before harvest have a white mildew growing on the skin.

The fungus over-winters in the soil or on old infected fruit, with the spores spread by wind, rain and insects. Continuous wet weather and temperatures of 22° to 25°C favour infection. It is suggested that growers clean up their orchard by removing shaded, infected and dead branches after harvest. Copper oxychloride during winter and copper sulphate in spring also help to reduce inoculum levels. These chemicals are replaced by fosetyl-Al or metalaxyl during flowering and fruit development.

Anthracnose

Anthracnose, *Colletotrichum gloeosporoides*, is a major disease in Guangdong, and also occurs in India. Although it attacks leaves and branches, along with flowers and flower stalks, infected fruit are unmarketable. Lesions on the leaves may appear as small round light grey areas, or irregular brown marks at the tips. In contrast, infections are much more obvious on the flowers and fruit. Outbreaks are common after warm wet weather. The fungus may not always cause immediate disease, which sometimes only becomes apparent after harvest. Fungicides are used during an initial outbreak, but are not always effective.

A form of anthracnose caused by *Colletotrichum gloeosporioides* (*Glomerella cingulata* in the sexual state) also affects trees in Australia. Pepper spot causes superficial skin blemishes to the fruit, but does not effect production, fruit quality or shelf life. More than half of the crop may be unmarketable in some orchards. The disease has been recorded on all cultivars, but is most severe on the popular "Kwai May Pink".

The disease first appears as brown pinhead freckles, usually on the top of semi-mature fruit, in areas with overhanging branches. The spots do not increase in size, but rapidly turn black. More spots appear on the top and sides of the fruit and may, by harvest, cover up to half of the fruit surface. Infections over-winter on the leaves, with the fungus potentially spread from nurseries to new orchards.

Until the appearance of pepper spot, lychee was generally free of diseases affecting fruit or foliage in Australia. However, the occurrence of the disease has resulted in attempts to control it using copper oxychloride and copper hydroxide. Calendar sprays of copper are costly and could lead to unacceptable residues if used close to harvest. Other chemicals such as mancozeb are being evaluated.

Tree Decline

A slow decline and a sudden death have been recorded in China, Viet Nam and Australia, especially in poorly drained soils. It can affect the whole tree or just one or two branches.

The symptoms include a sudden branch wilt that is followed by the decline of new growth on the affected branch over a period. In other situations, the tips die without wilting. The tree or branch may recover temporarily, but subsequently dies. Parts of the tree flush and grow, while other sections die.

A number of organisms including *Phytophthora*, *Pythium* and *Fusarium* have been isolated from the roots of trees, but it is not known where they cause the disease. A root rot caused by *Clitocybe* is reported to kill trees in the Philippines. Growers are advised not to plant on waterlogged soils.

Armillaria occasionally attacks roots and the base of trees of any age causing death or slow decline. The fungus may survive in the soil, or on stumps and roots of various trees for many years. The planting sites need to be fumigated before establishing new trees in the orchard.

Parasitic Algae and Nematodes

A parasitic algae, *Cephaleuros virescens*, occasionally attacks trees in Australia causing loss of vigour. Cultivars such as "Souey Tung" and "Haak Yip" are very susceptible. It can be controlled with two sprays of copper, before and after the wet season.

Nematodes such as *Xiphinema*, *Paratrichodorus* and *Helicotylenchus* are a problem in South Africa, but whether they are significant in Australia and Asia is not yet clear. Post-plant nematicides are used in South Africa, but have not been evaluated elsewhere.

Food Uses

Longans are much eaten fresh, out-of-hand, but some have maintained that the fruit is improved by cooking. In China, the majority are canned in sirup or dried. The canned fruits were regularly shipped from Shanghai to the United States in the past. Today, they are exported from Hong Kong and Taiwan.

For drying, the fruits are first heated to shrink the flesh and facilitate peeling of the rind. Then the seeds are removed and the flesh dried over a slow fire. The dried product is black, leathery and smoky in flavour and is mainly used to prepare an infusion drunk for refreshment.

A liqueur is made by macerating the longan flesh in alcohol.

Mamoncillo (Genip)

Melicoccus bijugatus, commonly called Spanish lime, genip, guinep, genipe, ginepa, quenepa, chenet, canepa, mamon, limoncillo or mamoncillo, is a fruit-bearing tree in the soapberry family Sapindaceae, native or naturalised over a wide area of the tropics including South and Central America, Puerto Rico, Dominican Republic, Haiti and other parts of the Caribbean, and parts of Africa and the Pacific.

The genus *Melicoccus* was first described by Patrick Browne, an Irish doctor and botanist, in 1756. This description was based on *M. bijugatus* trees which were cultivated in Puerto Rico . In 1760, Nikolaus Joseph von Jacquin described the first species in Browne's genus, which he named *M. bijugatus*. In 1762 Linnaeus used a spelling variation of the name *Melicocca bijuga*. Over the next two centuries, Linnaeus' spelling variation was used in almost all publications. A proposal was made in 1994 to conserve *Melicocca* over *Melicoccus*, but the proposal was rejected, leading to a restoration of the original version of the name.

In 1888 German taxonomist Ludwig Radlkofer placed *Melicoccus* in the tribe Melicocceae together with eight other genera. In his monograph on the Neotropical members of the tribe (*Talisia* and *Melicoccus*) Pedro Acevedo-Rodríguez suggested that although *Talisia* and *Melicoccus* appeared to form a monophyletic group, the other (Old World) genera probably did not belong to the same lineage.

The specific epithet *bijugatus* refers to the bijugate leaves, leaves which consist of two pairs of leaflets.

Local names

It is known by many names around the region: as *mamoncillo* or *mamón* (in Cuba, Costa Rica, Colombia, El Salvador, Nicaragua, Panama, and Venezuela). The fruit is called *chenette* (in Trinidad and Tobago), *gnep, guinep* (in Jamaica, the Cayman Islands, the United States Virgin Islands and Antigua and Barbuda), *guaya, quenepa* (in México), *skinnip* (in St. Kitts, "skinup" in (Grenada), *kenip* (in Dominica), *canepa, genip, guinep, ginepa, ginnip, kenèp, limoncillo* (in the Dominican Republic, Guyana, Haiti, Belize, Bahamas, Anguilla) and in some parts of Central

America *talpa jocote* (in some parts of Guatemala), *quenepa* (in Puerto Rico), *genepa, xenepa, kenepa* (in Curaçao and Aruba), *knippa* (in Suriname) and *Spanish lime* and *limoncillo* (in the Dominican Republic). Also, it is often referred to as *anoncillo* in central Cuba and southern Florida. It is called "ackee" in the countries of Barbados, St.Lucia, and St. Vincent and the Grenadines, however, in the rest of the Caribbean, the latter name is used to refer to the related *Blighia sapida*. In Grenada it is also called "Skin-Up."

Distribution

Melicoccus bijugatus is native to northern South America and naturalised in coastal and dry forest in Central America, the Caribbean and parts of the Old World tropics. It is believed to have been introduced into the Caribbean in pre-Columbian times. This fruit, known as *quenepa* in Puerto Rico, grows particularly abundant in the municipality of Ponce, and there is a yearly celebration in that municipality known as Festival Nacional de la Quenepa (National Genip Fruit Festival).

Description

Trees can reach heights of up to 25 m and come with alternate, compound leaves. The leaves have 4 elliptic leaflets which are 5-12.5 cm long and 2.5–5 cm (1-2 in.) wide. They are typically dioecious plants however polygamous trees occur f rom time to time. Flowers have 4 petals and 8 stamens and produce void, green drupes which are 2.5–4 cm long and 2 cm wide. Their pulp is orange, salmon or yellowish in colour with a somewhat juicy and pasty texture.

This fruit can be sweet or sour. In the southern areas of Mexico, it is generally eaten with chili powder, salt, and lime. The sweet varieties are generally eaten without condiments of any kind.

Cultivation

Being tropical, *M. bijugatus* prefers warmer temperatures. Its leaves can be damaged if the temperature hits the freezing point, with serious damage occurring below -4°C.

It is grown and cultivated for its ovoid, green fruit, which grow in bunches. The fruit, somewhat like a cross between a lychee and a lime, has a tight and thin, but rigid layer of skin, traditionally cracked by the teeth. Inside the skin is the tart, tangy, creamy pulp (technically the seed coat, or aril), which is sucked by putting the whole fruit inside the mouth (hence the name *mamoncillo* as *mamar* means "to suck") because the seed takes most of the volume of what is inside the skin. Despite the light colour of the fruit's flesh, the juice stains a dark brown colour, and was often used by indigenous Arawak natives to dye cloth. The species is also commonly planted along roadsides as an ornamental tree.

Varieties

Little horticultural attention has been given this fruit. In the 1950's, a large-fruited, sweet type was found in Key West. Air-layers and inarchings were made in order to permit trial of this type on the mainland. In the 1960's, horticulturist George Jackson evaluated the fruits of 54 trees in southern Puerto Rico. Fruits with less than 45% edible pulp and 20% total sugars were disregarded. He rated 9 trees as meriting further testing. Of these, 4 were selected as having the most desirable qualities. Their main characters were listed as follows:

'Puerto Rico #1'–round, of medium size, 28 to the lb (62/kg); flesh firm, semi-dry, separating easily from the seed; sweet, with 26.0% sugars.

'Puerto Rico #2'–round, of medium size, 27 to the pound (60/kg); rind medium-thick; flesh firm, semi-dry, separating easily from seed; sweet, 24.1% total sugars.

'Puerto Rico #3'–round -oblong, small, 49 to the pound (108/kg); rind thin, pliable; flesh firm, semi-dry, separating easily from the seed; very sweet, 24.1% total sugars.

'Puerto Rico #4'–round, medium-small, 40 to the lb (88/kg); rind medium-thin, flesh firm, semi-dry, separating easily from seed; agreeably acid and slightly sweet; 22.7% total sugar.

The percentage of edible matter by fruit-weight ranged from 46.6% to 48.6%.

In 1976, Dr. Carl Campbell of the University of Florida's Agricultural Research and Education Centre in Homestead, Florida, reported on his comparison of 3 selections made by interested individuals and an ordinary seedling growing at the Centre. The latter, labelled 'No. 1', was graded as: small, 49.1% pulp, but of only fair flavour, and poor annual yield.

'No. 2', or 'Queen', brought by W.F. Whitman from Key West; large, 55.6% pulp, only fair in flavour, and medium in yield.

'No. 3', brought by R.G. Newcomb from Key West; of good size, 48.2% pulp; of good flavour and borne heavily in most years.

'No. 4', or 'Montgomery', from the Montgomery (later, Jennings) Estate in Coral Gables; large, with sometimes 18% of crop having 2 seeds; 51.5% pulp; of good flavour, and borne heavily in most years.

Pollination

Generally, the presence of a male tree is necessary to pollinate the flowers of trees that are predominantly female (or hermaphrodite functioning as female). However, in Cuba, some trees have sufficient numbers of flowers of both sexes to yield regularly large crops without interplanting.

Climate

The mamoncillo is not strictly tropical, for it ascends up to 3,300 ft (1,000 m) above sea-level in South America. It can stand several degrees of frost in Florida. Nevertheless, it is too tender to fruit in California though it has been planted there on various occasions. It is well adapted to areas of low rainfall. That of Key West ranges from 30 to 50 in (75-125 cm) annually. The tree can tolerate long periods of drought.

Soil

In Cuba, the tree is said to flourish in nearly all types of terrain but particularly in deep, rich soil of calcareous origin. It seems perfectly at home in the oolitic-limestone of southern Florida and the Florida Keys. In Colombia, it has been observed to grow on such poor soils that it has been adopted for planting in soil reclamation efforts. It is spontaneous especially in dry, coastal districts.

Propagation

The mamoncillo is usually grown from seed but superior types should be vegetatively reproduced. Air-layering of fairly large branches, at least 2 in (5 cm) in diametre, is successful in the summer and there will be adequate root development in 5 to 6 weeks. Approach-grafting is feasible provided the rootstocks are raised in a lightweight medium, in plastic bags to facilitate attachment to the selected tree. Attempts to veneer-graft or chip-bud have generally failed.

Culture

Ordinarily, the mamoncillo, tree is given no care except for watering and fertilizing when first planted. Vegetatively propagated trees bear earlier than seedlings.

Season and Harvesting

In Florida, the fruits ripen from June to September. In the Bahamas, the season extends from July to October. Ladders and picking poles equipped with cutters are necessary in harvesting fruits from tall trees. The entire cluster is clipped from the branch when sampling indicates that the fruits are fully ripe. At this stage, the rind becomes brittle but does not change colour. If picked prematurely, the rind turns blackish, a sign of deterioration.

Keeping Quality

Because of the leathery skin, the fruit remains fresh for a long time and ships and markets well. The tropical horticulturist, David Sturrock, related that horsemen in Cuba often hung branches of mamoncillos on the saddle horn to enjoy and relieve thirst during long rides.

Pests and Diseases

The tree is a host of the Citrus black fly, *Aleurocanthus woglumi*. There are several parasites (*Prospaltella* spp., *Eretmocerus serius*, and *Amitus hesperidium*) which provide effective control of this pest. In Florida, *Armillariella (Clitocybe) tabescens* causes mushroom root rot; *Fusarium* and *Phyllosticta* cause leaf spot; and *Cephaleuros virescens*, algal leaf spot and green scurf.

Mango

The mango is a juicy stone fruit belonging to the genus *Mangifera*, consisting of numerous tropical fruiting trees, that are cultivated mostly for edible fruits. The majority of these species are found in nature as wild mangoes. They all belong to the flowering plant family Anacardiaceae. The mango is native to South and Southeast Asia, from where it has been distributed worldwide to become one of the most cultivated fruits in the tropics. The highest concentration of Mangifera genus is situated in western part of Malesia (Sumatra, Java and Borneo) and in Burma and India. While other *Mangifera* species (e.g. horse mango, *M. foetida*) are also grown on a more localized basis, *Mangifera indica* – the 'common mango' or 'Indian mango' – is the only mango tree commonly cultivated in many tropical and subtropical regions. It originated in India and Burma. It is the national fruit of India and the Philippines and the national tree of Bangladesh. In several cultures, its fruit and leaves are ritually used as floral decorations at weddings, public celebrations and religious ceremonies.

Mango trees (*Mangifera indica* L.) grow up to 35–40 m (115–131 ft) tall, with a crown radius of 10 m (33 ft). The trees are long-lived, as some specimens still fruit after 300 years. In deep soil, the taproot descends to a depth of 6 m (20 ft), with profuse, wide-spreading feeder roots; the tree also sends down many anchor roots, which penetrate several feet of soil. The leaves are evergreen, alternate, simple, 15–35 cm (5.9–13.8 in) long and 6–16 cm (2.4–6.3 in) broad; when the leaves are young they are orange-pink, rapidly changing to a dark, glossy red, then dark green as they mature. The flowers are produced in terminal panicles 10–40 cm (3.9–15.7 in) long; each flower is small and white with five petals 5–10 mm (0.20–0.39 in) long, with a mild, sweet odor suggestive of lily of the valley. There are over 400 varieties of mango, many of which ripen in summer while some give double-crop. The fruit takes three to six months to ripen.

The ripe fruit varies in size and colour. Cultivars are variously yellow, orange, red or green, and carry a single flat, oblong pit that can be fibrous or hairy on the surface, and which does not separate easily from the pulp. Ripe, unpeeled mangoes give off a distinctive resinous, sweet smell. Inside the pit 1–2 mm (0.039–0.079 in) thick is a thin lining covering a single seed, 4–7 mm (0.16–0.28 in) long. The seed contains the plant embryo.

Cultivation

Mangoes have been cultivated in South Asia for thousands of years and reached East Asia between the fifth and fourth centuries BC. By the 10th century AD, cultivation had begun in East Africa. The 14th century Moroccan traveller, Ibn Battuta, reported it at Mogadishu. Cultivation came later to Brazil, the West Indies and Mexico, where an appropriate climate allows its growth.

The mango is now cultivated in most frost-free tropical and warmer subtropical climates; almost half of the world's mangoes are cultivated in India alone, with the second-largest source being China. Mangoes are also grown in Andalusia, Spain (mainly in Málaga province), as its coastal subtropical climate is one of the few places in mainland Europe that allows the growth of tropical plants and fruit trees. The Canary Islands are another notable Spanish producer of the fruit. Other cultivators include North America (in South Florida and California's Coachella Valley), South and Central America, the Caribbean, Hawai'i, south, west and central Africa, Australia, China, Pakistan, Bangladesh, and Southeast Asia. Though India is the largest producer of mangoes, it accounts for less than one percent of the international mango trade; India consumes most of its own production.

Many commercial cultivars are grafted on to the cold-hardy rootstock of *Gomera-1* mango cultivar, originally from Cuba. Its root system is well adapted to coastal Mediterranean climate. Many of the 1,000+ mango cultivars are easily cultivated using grafted saplings, ranging from the "turpentine mango" (named for its strong taste of turpentine) to the *huevos de toro*. Dwarf or semi-dwarf varieties serve as ornamental plants and can be grown in containers. A wide variety of diseases can afflict mangoe.

Potential for Contact Dermatitis

Contact with oils in mango leaves, stems, sap, and skin can cause dermatitis and anaphylaxis in susceptible individuals. It contains mangiferen, resinous acid, mangiferic acid, and the resinol called mangiferol. Those with a history of poison ivy or poison oak contact dermatitis may be most at risk for mango contact dermatitis. Cross-reactions between mango allergens and urushiol, a chemical in poison ivy and poison sumac that can cause dermatitis, have been observed. Urushiol is also present in mango leaves and stems. During its primary ripening season, it is the most common cause of plant dermatitis in Hawaii. After contacting it, reactions may not be immediate. Eyelids, face, or other parts of the body may even swell because of this. It irritates the skin and may even blister the skin. Also, burning of the mango wood, leaves, etc. should be avoided because fumes could be dangerous.

Cultivars

Many hundreds of named mango cultivars exist. In mango orchards, several cultivars are often crossed to improve pollination. Many desired cultivars are monoembryonic and must be propagated by grafting or they do not breed true. A common mono-embryonic cultivar is Alphonso, an important export product, considered as "the king of mangoes".

Cultivars that excel in one climate may fail elsewhere. For example, Indian cultivars such as Julie, a prolific cultivar in Jamaica, require annual fungicide treatment to escape a lethal fungal disease known as anthracnose in Florida. Asian mangoes are resistant to anthracnose.

The current world market is dominated by the cultivar Tommy Atkins, a seedling of Haden that first fruited in 1940 in southern Florida, U.S. It was initially rejected commercially by Florida researchers. For example, 80% of mangoes in UK supermarkets are Tommy Atkins. Despite its fibrous flesh and only fair taste, growers worldwide have embraced the cultivar for its exceptional productivity and disease resistance, shelf life, transportability, size and appealing colour.

Alphonso, Benishaan and Kesar mango varieties are popular varieties in India's southern states, while the Chaunsa variety, among others, is popular in the northern states and Pakistan.

Guatemala markets sell a variety called 'mango de leche' which is more resinous outside and inside.

Generally, ripe mangoes have an orange-yellow or reddish peel and are juicy for eating, while exported fruit are often picked while underripe with green peels. Although producing ethylene while ripening, unripened exported mangoes do not have the same juiciness or flavour as fresh fruit.

Like other drupaceous fruits, mangoes come in both freestone and clingstone varieties.

Mango Pests

The following are some of the high priority exotic pests of mangoes, as identified in the development of the *Industry Biosecurity Plan for the Mango Industry*.

Any of these pests would have serious consequences should they enter and become established in Australia. Additional information is included in the fact sheets. Implementing biosecurity measures to control endemic pests will go a long way towards preventing exotic pests from entering and becoming established on your farm.

Oriental Fruit Fly Complex

- Can cause up to 100% fruit loss
- Adults 6-8 mm long with a narrow brown band along the edge of the wings
- Abdomen has a distinctive black T-shaped mark, which is similar to a number of other endemic species
- Larval feeding can result in rotting of fruit and may cause fruit to drop
- Long range dispersal through movement of larvae-infested fruit

Mango Pulp Weevil

- Causes high economic losses to mango industries in Asia
- Adults are 6-9 mm long, stout, and dark brown with paler patches
- Unlike the Mango seed weevil, larvae only feed on the pulp
- Circular exit holes appear in the fruit when adult weevils emerge
- Long range dispersal through movement of infested fruit

Mango Malformation Disease

- Found in India, Egypt, Africa, Central America, Brazil, Mexico, Malaysia, Pakistan
- Most important floral disease of mangoes worldwide

- Causes malformed shoots and inflorescences
- Panicles are thickened with large flowers and inflorescences appear crowded
- Long range dispersal via infected plant material and pruning equipment

Mango Gall Midges

- Found in India, Africa, Malaysia
- Eggs laid on leaves cause small red spots
- Wart-like galls form quickly and extend up to 3-4 mm in diametre
- Heavily galled leaves can curl up and drop off prematurely, causing dieback of whole branches in susceptible cultivars
- Long range dispersal through movement of infested plant material

Red Banded Mango Caterpillar

- Found in Southeast Asia, Papua New Guinea, Torres Strait Islands and quarantined on the tip of Cape Peninsula
- Larvae have distinctive red and white bands plus a black collar
- Damaging to all stages of fruit development, feeding on both flesh and seed
- Caterpillar entry holes lead to sap stain on fruit
- Long range dispersal through movement of infested plant material

Blue-striped Nettle Grub

- Found in Vietnam, Indonesia, China, Papua New Guinea
- Larvae have thick, green fleshy bodies with three pale blue longitudinal stripes
- Cocoons are buried in soil or among leaf litter
- Grubs chew holes in leaves; larger infestations lead to defoliation of entire trees
- Long range dispersal through movement of infested plant material

Mango Fruit Borer

- Found in India, Indonesia and parts of the Northern Territory
- Larvae are reddish violet, and turn dark blue as they develop
- Larvae feed on pulp
- Causes premature fruit drop, particularly in young fruit
- Long range dispersal through movement of infested fruit

Mango Sudden Decline Syndrome

- Found in Brazil, Pakistan, Oman
- Symptoms vary depending on which species of Ceratocystis is present
- Causes parts or all of the tree to wilt and die
- Blue-grey to brown staining of vascular tissue
- Lifecycle not well understood but likely spread with the Mango bark beetle

Pitaya (Dragon Fruit)

A pitaya or pitahaya is the fruit of several cactus species. "Pitaya" usually refers to fruit of the genus *Stenocereus*, while "Pitahaya" or "Dragonfruit" always refers to fruit of the genus *Hylocereus*.

These fruits are commonly known in English as "dragon fruit", reflecting its vernacular Asian names. These include the Malay *buah naga* (lit. dragon fruit), the Khmer *sror kaa neak* (dragon scale), the Thai *kaeo mangkon* (dragon crystal), the Lao *maak manggohn*, the Vietnamese *thanh long* (green dragon), and the Chinese *huÒ lóng guÒ* (fire dragon fruit) or *lóng zhû guÒ* (dragon pearl fruit). Other vernacular names are strawberry pear or nanettika fruit. Pitahaya-producing cacti of the genus *Hylocereus* are originally native to Mexico, and they were also transplanted to Central America and to other parts of the world. Currently, they are also cultivated in East Asian and Southeast Asian countries such as Cambodia, Thailand, Taiwan, Malaysia, Vietnam, Sri Lanka, the Philippines, Indonesia and more recently Bangladesh. They are also found in Okinawa, Hawaii, Israel, northern Australia, southern China and in Cyprus.

Figure: Stenocereus queretaroensis *pitaya prepared for eating*

The fruit was probably introduced by Europeans who brought it from the New World. In the case of Taiwan, the fruit was brought in by the Dutch.

Hylocereus blooms only at night; the large white fragrant flowers of the typical cactus flower shape are among those called "moonflower" or "Queen of the Night". Sweet pitahayas have a creamy pulp and a delicate aroma. It is also grown as an Ornamental plant, used in gardens as a flowering vine and a house plant indoors.

Pitaya Varieties

Stenocereus fruit (sour pitayas) are a variety that is commonly eaten in the arid regions of the Americas. They are more sour and refreshing, with juicier flesh and a stronger taste. The sour pitaya or pitaya agria (*S. gummosus*) in the Sonoran Desert has been an important food source for Native Americans. The Seri people of northwestern Mexico still harvest the fruit, and call the plant *ziix is ccapxl* – "thing whose fruit is sour". The fruit of related species, such as *S. queretaroensis* and the dagger cactus (*S. griseus*), are also locally important food. The Organ Pipe Cactus (*S. thurberi*) fruit (called *ool* by the Seris) is the *pitaya dulce* ("sweet pitaya"). It still has a more tart aroma than *Hylocereus* fruit, described as somewhat reminiscent of watermelon; it has some uses in folk medicine.

Fruits of some other columnar cacti (mainly Cereeae) are also called "pitayas" – for example those of the Peruvian Apple Cactus (*Cereus repandus*), which are very rare.

Sweet pitayas come in three types, all with leathery, slightly leafy skin:

- *Hylocereus undatus* (Pitaya blanca or White-fleshed Pitaya) has red-skinned fruit with white flesh. This is the most commonly seen "dragon fruit".
- *Hylocereus costaricensis* (Pitaya roja or Red-fleshed Pitaya, also known as *Hylocereus polyrhizus*) has red-skinned fruit with red flesh.
- *Hylocereus megalanthus* (Pitaya amarilla or Yellow Pitaya, also known as *Selenicereus megalanthus*) has yellow-skinned fruit with white flesh.

Early imports from Colombia to Australia were designated *Hylocereus ocampensis* (supposedly red fruit) and *Cereus triangularis* (supposedly yellow fruit). It is not quite certain to which species these taxa refer, though the latter is probably the red pitahaya.

The fruit can weigh from 150 to 600 grams; some may reach one kilogram.

Cultivation

Figure: *Cereus peruvianus (Cereus repandus) Pitaya plants in Sde Nitzan (Israel)*

After thorough cleaning of the seeds from the pulp of the fruit, the seeds may be stored when dried. Ideally, the fruit must be unblemished and overripe. Seeds grow well in a compost or potting soil mix - even as a potted indoor plant. Pitaya cacti usually germinate between 11 and 14 days after shallow planting. As they are cacti, overwatering is a concern for home growers. As their growth continues, these climbing plants will find something to climb on, which can involve putting aerial roots down from the branches in addition to the basal roots. Once the plant reaches a mature 10 pounds in weight, the plant may flower.

Pitaya flowers bloom overnight and usually wilt by the morning. They rely on nocturnal pollinators such as bats or moths for fertilization. Self-fertilization will not produce fruit in some species, and while cross-breeding has resulted in several "self-fertile" varieties, cross-pollinating with a second plant species generally increases fruit set and quality. This limits the capability of home growers to produce the fruit. However, the plants can flower between three and six times in a year depending on growing conditions. Like other cacti, if a healthy piece of the stem is broken off, it may take root in soil and become its own plant.

The plants can handle temperatures up to 40 °C (104 °F) and very short periods of frost, but will not survive long exposure to freezing temperatures. The cacti thrive most in USDA zones 10-11, but may survive outdoors in zone 9a or 9b.

Hylocereus has adapted to live in dry tropical climates with a moderate amount of rain. The dragon fruit sets on the cactus-like trees 30–50 days after flowering and can sometimes have 5-6 cycles of harvests per year. There are some farms in Vietnam that produce 30 tons of fruit per hectare every year.

Pests and diseases

Overwatering or excessive rainfall can cause the flowers to drop and fruit to rot. Also, extended over-watering can cause maturing fruit to split on the branch. Birds can be a nuisance. The bacterium *Xanthomonas campestris* causes the stems to rot. *Dothiorella* fungi can cause brown spots on the fruit, but this is not common.

Consumption

To prepare a pitaya for consumption, the fruit is cut open to expose the flesh. The fruit's texture is sometimes likened to that of the kiwifruit because of its black, crunchy seeds. The flesh, which is eaten raw, is mildly sweet and low in calories. The seeds are eaten together with the flesh, have a nutty taste and are rich in lipids, but they are indigestible unless chewed. The fruit is also converted into juice or wine, or used to flavour other beverages. The flowers can be eaten or steeped as tea. The skin is not eaten, and in farm-grown fruit it may be polluted with pesticides. Ingestion of significant amounts of red-fleshed dragon fruit (such as Costa Rican Pitaya) may result in a harmless reddish colouration of the urine (pseudohematuria) and of the feces.

Several of the Padres who missionized Baja California recorded an unusual form of consumption of pitaya that is also shared in some O'odham stories from southern Arizona. It is called the "second harvest" of pitaya seeds. With the scarcity of fruits in their lands, the pitaya was such a prized fruit that once it was eaten, the natives would wait for their own excrement to dry, then break it apart separating the pitaya seeds. These seeds would be ground into a flour and eaten again, giving the pitaya's "second harvest" its name. Interestingly, the O'odham name for the Milky Way translates as "the second harvest of pitaya."

Taste

The mild taste of pitaya flesh is often remarked upon, as it stands in stark contrast to the vibrant exterior. The taste has been described as being "very bland... like a melon or kiwi," with a "mild sweetness."

Sapodilla

Manilkara zapota, commonly known as the sapodilla, is a long-lived, evergreen tree native to southern Mexico, Central America and the Caribbean. An example natural occurrence is in coastal Yucatán in the Petenes mangroves ecoregion, where it is a subdominant plant species. It was introduced to the Philippines during Spanish colonization. It is grown in large quantities in India, Thailand, Malaysia, Cambodia, Indonesia, Bangladesh and Mexico.

Sapodilla can grow to more than 30 m (98 ft) tall with an average trunk diametre of 1.5 m (4.9 ft). The average height of cultivated specimens, however, is usually between 9 and 15 m (30 and 49 ft) with a trunk diametre not exceeding 50 cm (20 in). It is wind-resistant and the bark is rich in a white, gummy latex called chicle. The ornamental leaves are medium green and glossy. They are alternate, elliptic to ovate, 7–15 cm long, with an entire margin. The white flowers are inconspicuous and bell-like, with a six-lobed corolla. An unripe fruit has a firm outer skin and when picked, releases white chicle from its stem. A fully ripened fruit has saggy skin and does not release chicle when picked.

The fruit is a large ellipsoid berry, normally 4–8 but up to 15 cm in diametre, containing two to five seeds. Inside, its flesh ranges from a pale yellow to an earthy brown colour with a grainy texture akin to that of a well-ripened pear. The seeds are black and resemble beans, with a hook at one end that can catch in the throat if swallowed.

The fruit has an exceptionally sweet, malty flavor. The unripe fruit is hard to the touch and contains high amounts of saponin, which has astringent properties similar to tannin, drying out the mouth.

The trees can only survive in warm, typically tropical environments, dying easily if the temperature drops below freezing. From germination, the sapodilla tree will usually take anywhere from five to eight years to bear fruit. The sapodilla trees yield fruit twice a year, though flowering may continue year round.

Growing sapodillas are not strictly tropical and adult sapodilla fruit trees can survive temperatures of 26-28 degrees F. for a short period of time. Sapling trees are more likely to sustain major damage or even die at 30 degrees F. Growing sapodillas are not particular when it comes to water requirements. They may do equally well in arid or humid environments, although more severe conditions may result in lack of fruiting.

Despite its temperature tolerance, if you want to grow a sapodilla tree in a less than semi-tropical area, it would be prudent to either grow it in a greenhouse or as a container plant that can be moved to a protected area in case of inclement weather. If such weather occurs, the tree may also be covered with sheeting to aid in protection.

This evergreen fruit bearer hails from the family of Sapotaceae in the genus of *Manilkara* with a calorie rich, easy-to-digest fruit. The sapodilla fruit is sand coloured with a skin similar to a kiwi but without the fuzz. The interior pulp is of young sapodilla fruit is white with a heavy concentration of sticky latex, called saponin. The saponin abates as the fruit ripens and the flesh subsequently turns brown. The inside of the fruit contains 3-10 inedible seeds at the centre.

A good reason to grow a sapodilla tree is its excellent source of nutrition within the fruit, which is composed of fructose and sucrose and is rich in calories. The fruit also contains vitamins such as vitamin C and A, folate, niacin and pantothenic acid and minerals like potassium, copper, and iron. It is rich in antioxidant tannins too and purported to be useful as an anti-inflammatory and a virus, "bad" bacteria and parasite fighter. Sapodilla fruit has also been used as an anti-diarrheal, hemostatic, and hemorrhoid aid.

Care for Sapodilla Trees

To grow a sapodilla tree, most propagation is done by seed, which will be viable for years although some commercial growers use grafting and other practices. Once germinated, use some patience as it takes 5-8 years to grow a sapodilla tree of bearing age.

As mentioned, the fruit tree is tolerant of most conditions but prefers a sunny, warm, and frost free location in most any type of soil with good drainage.

Additional care for sapodilla trees advises fertilizing the young trees with -8% nitrogen, 2-4% phosphoric acid and 6-8% potash every

2 or 3 months with ¼ pound and increasing gradually to 1 pound. After the first year, 2 or 3 application a year is plenty.

Not only are sapodilla trees tolerant of drought conditions, but they can take soil salinity, need very little pruning and are mostly pest resistant.

As long as the sapodilla tree is protected from frost and patience is in abundance for this slow grower, flavorful fruit shall be the reward from this tolerant specimen.

Black Sapote

Diospyros nigra, the Black Sapote, is a species of persimmon that is native to eastern Mexico and Central America south to Colombia. Other names include Chocolate Pudding Fruit, Chocolate Persimmon and (in Spanish) *Zapote Prieto*.

Mature trees can grow to over 25 m (82 ft) in height and are evergreen. It is frost sensitive but can tolerate light frosts. The leaves are elliptic-oblong, tapered at both ends, dark green, glossy and 10–30 cm (3.9–11.8 in) long. Some trees bear only male flowers. Others have both male and female flowers, though some of these are self-incompatible. Fruiting takes about 3–4 years from seed and the trees are heavy bearers.

Fruit

Black Sapote fruit are tomato-like and measure 5–10 cm (2.0–3.9 in) in diametre, with an inedible skin that turns from olive to a deep yellow-green when ripe and a pulp which is white and inedible when unripe but assumes a flavor, colour and texture often likened to chocolate pudding when ripe. Fruits usually contain seeds, up to a maximum of 12. The texture has been likened to that of a papaya. Boning (2006), describes the ripe fruit as having "the taste and consistency of chocolate pudding."

Unripe fruits are astringent, caustic, bitter, and an irritant.

Propagation

Propagation is usually from seed, which can retain viability for several months and require around 30 days for germination. Some trees are seedless however, and can be propagated by air-layering or shield-budding.

Cultivation

Black Sapote trees are normally found below 600 metres, but are not particular about soil, and can tolerate light frosts. They are sensitive to drought, requiring irrigation in dry areas, but are quite tolerant of flooding. The tree grows fairly slowly for the first 3–4 years, perhaps just 1 foot/year for the first couple of years. Later however it grows much more rapidly. Trees should be spaced 10-12m apart.

Cultivars

The range in size of tree and hairiness of leaves; size, shape, seediness, flesh colour and sweetness of fruit; and time of fruiting suggest that considerable genetic variability exists. Selections have been made and propagated in the Philippines, Australia, and Florida, USA.

Mamey Sapote

Pouteria sapota, mamey sapote, is a species of tree native to Central America, naturally ranging from southern Mexico to southern Costa Rica. Today, the tree is cultivated not only in Mexico, but also in Central America, the Caribbean, and South Florida for its fruit, which is commonly eaten in many Latin American countries. It has different names depending on the country: mamey, *zapote colorado* (Costa Rica), *níspero* and *zapote rojo* (South America), among others.

Mamey sapote is a large and highly ornamental evergreen tree that can reach a height of 15 to 45 metres (60 to 140 feet) at maturity. Like most fruit trees, it is mainly propagated by grafting, which ensures the new plant has the same characteristics as the parent, especially its fruit. It is also considerably faster than growing trees by seed.

The leaves are pointed at both ends, 4 to 12 inches in length and grow in clusters at the ends of branches.

The fruit is about 10 to 25 cm (4 to 10 inches) long and 8 to 12 cm (3 to 5 inches) wide and has flesh ranging in colour from pink to orange to red. The brown skin has a texture somewhat between sandpaper and the fuzz on a peach. The fruit's texture is creamy and soft. A mamey sapote is ripe when the flesh is pink when a fleck of the skin is removed. The flesh should give slightly, as with a ripe kiwifruit.

The mamey sapote is related to other sapotes such as *sapodilla* (*Manilkara zapota*), abiu (*P. caimito*) and canistel (*P. campechiana*), but unrelated to the black sapote (*Diospyros digyna*) and white sapote (*Casimiroa edulis*). It should not be confused with the mammee apple (*Mammea americana*).

Uses

The fruit is eaten raw or made into milkshakes, smoothies, ice cream and fruit bars. It can be used to produce marmalade and jelly.

Some consider the fruit to be an aphrodisiac. Some beauty products use oil pressed from the seed, otherwise known as sapayul oil.

Nutrition

The fruit is an excellent source of vitamin B6 and vitamin C, and is a good source of riboflavin, niacin, vitamin E, manganese, potassium and dietary fibre. Research has identified several new carotenoids from the ripe fruit.

Caring for a Mamey Tree

Step 1: Plant your young tree in deep, well-drained soil and full sun at the beginning of your rainy season. Dig a planting hole that is three to four times larger than the tree's root ball and then refill it with some of the soil you dug out, providing enough space for the tree to fill the hole up to the top of its root system. It is not necessary to dig in topsoil, compost or fertilizer. Be sure to leave 20 or 30 feet between trees if you are planting more than one, because it will grow quite large.

Step 2: Water your newly planted mamey immediately after planting and every day for its first 6-8 weeks in the ground. Give mature trees one inch of irrigation once or twice a week if rains do not provide this amount.

Step 3: Spread a 2- to 6-inch layer of organic mulch around the base of the tree. Bark, wood chips or compost are recommended as mulch for the mamey.

Step 4: Prune suckers from your mamey tree during summer to encourage the strong growth of its main trunk. On mature trees, always prune dead and diseased branches back to the main trunk. You can keep the mamey at 12 or 15 feet tall in order to enable easier harvesting of the fruit and to help prevent upper branches from breaking during high winds.

Step 5: Fertilize your mamey with a balanced plant food when you first see signs of new growth and again during its active growing season, which is March through October in most tropical regions. After the tree is 3 years old, reduce the amount of fertilizer you use to half the original application and feed it only at the beginning of each growing season.

White Sapote

White sapote, *Casimiroa edulis*, also known as *cochitzapotl* in the Nahuatl language (meaning "'sleep-sapote") is a species of tropical fruiting tree in the family Rutaceae, native to eastern Mexico and Central America south to Costa Rica.

Mature *Casimiroa edulis* trees range from 5–16 metres (16–52 ft) tall and are evergreen. The leaves are alternate, palmately compound with 3–5 leaflets, the leaflets 6–13 cm long and 2.5–5 cm broad with an entire margin, and the leaf petiole 10–15 cm long.

The fruit is an ovoid drupe, 5–10 cm in diametre, with a thin, inedible skin turning from green to yellow when ripe, and an edible pulp, which can range in flavor from bland to banana-like to peach to pear to vanilla flan. The pulp can be creamy-white in green skin varieties or a beige-yellow in yellow skin varieties and has a smooth texture similar to ripe avocado. It contains from one to five seeds that are said to have narcotic properties.

Chemical Constituents

In the past 40 years, experiments have been carried out on the white sapote's seeds that have yielded the identity of many pharmacologically active compounds, including: *N*-methylhistamine, *N*,*N*-dimethylhistamine and histamine. It also contains 22 ,5,6-Trimethoxyflavone, 22 ,5,6,7-tetramethoxyflavone (zapotin) and 5-hydroxy-22 ,6,7-trimethoxyflavone (zapotinin).

Health Effects

Several recent *in vitro* studies have shown that zapotin has potential anti-carcinogenic effects against isolated colon cancer cells.

Eating the fruit has long been known to produce drowsiness, as noted by Francisco Hernández de Toledo in the 16th century. Europeans were not the first to note this quality; the Nahuatl name of the plant itself is a record of the fact that people knew the plant induced drowsiness.

Taxonomy

Unlike the mamey sapote, white sapote is a member of the family Rutaceae, to which citrus belongs. The black sapote is also unrelated and is actually a species of persimmon. This confusion may be because "sapote" comes from the Nahuatl (Aztec) word *tzapotl*, used to describe all soft, sweet fruit. Commonly grown in Northern New South Wales, Australia, and often mistaken for a persimmon, although these two fruits are unrelated.

Climate

White sapote trees are best adapted to subtropical climates with a distinct cool and or dry period. Trees do produce well in the dry tropics or those areas with a distinct dry period. A period of environmentally induced (i.e., cool temperatures and/or dry period) dormancy leads to flowering once warm temperatures and/or the rainy season commences.

White sapote trees are moderately tolerant of freezing temperatures with young trees damaged or killed at temperatures at or below 24°F (-4°C) and mature trees damaged or killed at temperatures at or below 26°F (-3°C).

Propagation

White sapote trees may be propagated by seed or vegetatively. Seeds should be planted within 3 weeks of harvesting from the fruit, and seedlings may begin to bear in 7 to 8 years. White sapote varieties do not come true to seed and therefore must be vegetatively propagated by grafting or budding onto seedling rootstock. Grafting and budding is most sucessful during the warm season when trees are actively growing.

Production (Crop Yields)

Although not extensively documented, white sapote trees are reported to be very productive. Trees generally flower from late fall to early summer. The harvest season varies with variety but generally is during the spring and summer.

Spacing

Although white sapote grows moderately slowly, trees that are not pruned eventually need a lot of space because they will develop a large canopy. White sapote trees in the home landscape should be planted 25 ft (7.6 m) or more feet away from the nearest tree and/or structure. Trees planted too close to other trees or structures may not grow normally or produce much fruit due to shading.

Soils

White sapote trees are well-adapted to most well-drained soils from sands, to clays, to limestone-based soils.

Planting a White Sapote Tree

Proper planting is one of the most important steps in successfully establishing and growing a strong, productive tree. The first step is to choose a healthy nursery tree. Commonly, nursery white sapote trees are grown in 3-gallon (11-litre) containers, and trees stand 2 to 4 ft (0.6-1.2 m) from the soil media. Large trees in smaller containers should be avoided because the root system may be "root bound." This means all the available space in the container has been filled with roots to the point that the tap root is growing along the edge of the container in a circular fashion. Root bound root systems may not grow properly once planted in the ground. Inspect the tree for insect pests and diseases, and inspect the trunk of the tree for wounds and constrictions. Select a healthy tree and water it regularly in preparation for planting in the ground.

Site Selection

In general, white sapote trees should be planted in full sun for best growth and fruit production. Select a part of the landscape away from other trees, buildings and structures, and power lines. Remember, sapote trees can become very large if not pruned to contain their size. Select the warmest area of the landscape that does not flood (or remain wet) after typical summer rains.

Planting in Sandy Soil

Many areas in Florida have sandy soil. Remove a 3- to 10-ft-diametre (0.9- to 3.1-m) ring of grass sod. Dig a hole 3 to 4 times the diametre and 3 times as deep as the container the white sapote tree came in. Making a large hole loosens the soil next to the new tree, making it easy for the roots to expand into the adjacent soil. It is not necessary to apply fertilizer, topsoil, or compost to the hole. In fact, placing topsoil or compost in the hole first and then planting on top of it is not desirable. If you wish to add topsoil or compost to the native soil, mix it with the excavated soil in no more than a 1:1 ratio.

Backfill the hole with some of the excavated soil removed. Remove the tree from the container and place it in the hole so that the top of the soil media from the container is level with or slightly above the surrounding soil level. Fill soil in around the tree roots and tamp slightly to remove air pockets. Immediately water the soil around the tree and tree roots. Staking the tree with a wooden or bamboo stake is optional. However, do not use wire or nylon rope to tie the tree to the stake because they may eventually damage the tree trunk as it grows. Use a cotton or natural fibre string that will degrade slowly.

Planting in Rockland Soil

Many areas in Miami-Dade County have a very shallow soil, and several inches below the soil surface is a hard calcareous bedrock. Remove a 3- to 10-ft-diametre (0.9- to 3.1-m) ring of grass sod. Make a hole 3 to 4 times the diametre and 3 times as deep as the container the white sapote tree came in. To dig a hole, use a pick and digging bar to break up the rock or contract with a company that has auguring equipment or a backhoe. Plant the tree as described for sandy soils.

Planting on a Mound

Many areas in Florida are within 7 ft or so of the water table and experience occasional flooding after heavy rains. To improve plant survival, consider planting fruit trees on a 2- to 3-ft-high by 4- to 10-ft-diametre (0.6- to 0.9-m by 1.2- to 3.1-m) mound of native soil. After the mound is made, dig a hole 3 to 4 times the diametre and 3 times as deep as the container the tree has come in. In areas where the bedrock nearly comes to the surface (rockland soil); follow the recommendations for the previous section. In areas with sandy soil, follow the recommendations from the section on planting in sandy soil.

Care of White Sapote Trees in the Home Landscape

A calendar outlining the month-to-month cultural practices for white sapote is shown.

Fertilizer

White sapote does not appear to be demanding in its fertilizer requirements. After planting, when new growth begins, apply 1/4 lb (113 g) of a young tree fertilizer such as a 6-6-6-2 (%nitrogen-% phosphate-% potash-% magnesium) with minor elements with 20 to 30% of the nitrogen from organic sources. Repeat this every 6 to 8 weeks for the first year, then gradually increase the amount of fertilizer to 0.5, 0.75, 1.0 lb (227 g, 341 g, 454 g) as the tree grows. Use 4 to 6 minor element (nutritional) foliar sprays per year from April to September.

White sapote trees generally do not develop iron deficiency, even when grown in the rocky, calcareous, high-pH soils of Miami-Dade County. If iron deficiency symptoms appear (chlorotic leaves with green veins) apply iron. For trees in acid to neutral soils, apply dry iron sulfate at 0.25 to 1 oz per tree (7-28 g) to the soil 2 to 4 timers per year; water the iron into the ground. In alkaline soils with a high-pH, drench the soil adjacent to the tree trunk with iron chelate 1 to 2 times per year from June through September. Summarizes the fertilizer recommendations for white sapote.

For mature trees, 2.5 to 5.0 lbs (1.1-2.3 kg) of fertilizer per application 2 to 3 times per year is recommended. The fertilizer mix should also include phosphate (P_2O_5) and potash (K_2O); use a 6-6-6, 8-3-9 or similar material. Use 2 to 3 minor element (nutritional) foliar sprays per year from April to September.

Irrigation (Watering)

Newly planted white sapote trees should be watered at planting and every other day for the first week or so, and then 1 to 2 times a week for the first couple of months. During prolonged dry periods (e.g., 5 or more days of little to no rainfall), newly planted and young white sapote trees (first 3 years) should be watered once a week. Once the rainy season arrives, irrigation frequency may be reduced or stopped. Once white sapote trees are 4 or more years old, watering will be beneficial to plant growth and crop yields only during very prolonged dry periods during the year. Mature white sapote trees do not need frequent watering and over watering may cause trees to decline or be unthrifty.

White Sapote Trees and Lawn Care

White sapote trees in the home landscape are susceptible to trunk injury caused by lawn mowers and weed eaters. Maintain a grass-free area 2 to 5 or more feet (0.6-1.5 m) away from the trunk of the tree. Never hit the tree trunk with lawn mowing equipment and never use a weed eater near the tree trunk. Mechanical damage to the trunk of the tree will weaken the tree, and if severe enough, can cause dieback or kill the tree.

Roots of mature white sapote trees spread beyond the drip-line of the tree canopy, and heavy fertilization of the lawn next to white sapote trees is not recommended, because it may reduce fruiting and or fruit quality. The use of lawn sprinkler systems on a timer may result in over watering and cause white sapote trees to decline. This is because too much water too often applied causes root rot.

Mulch

Mulching white sapote trees in the home landscape helps retain soil moisture, reduces weed problems next to the tree trunk, and improves the soil near the surface. Mulch with a 2- to 6-inch (5- to 15-cm) layer of bark, wood chips, or similar mulch material. Keep mulch 8 to 12 inches (20-30 cm) from the trunk.

Insect Pests

There are few insect problems, although some varieties may be susceptible to fruit fly attack. Scales occasionally attack leaves. Contact your local County Extension Agent for current control recommendations.

Diseases

Some varieties are susceptible to scab, which causes irregularly-shape brown corky areas on the peel surface that usually split. Contact your local County Extension Agent for current control recommendations.

Pruning

Formative pruning during the first 2 years may be desirable to encourage lateral branching and growth. After several years of production, it is desirable to cut back the tops of the trees to 10 to 15 feet (3.1 to 4.6 m). Selectively removing a few upper limbs back to their origins (crotches) each year will help prevent the loss of the lower tree canopy due to shading by the upper canopy. In addition, maintaining a smaller tree facilitates tree care and fruit harvest, makes it easier to spray the tree, and greatly reduces possible storm damage. Do not remove lower tree branches.

Pruning should be done soon after harvest. Annual or biannual selective pruning can control the spread and limit tree height to 10- to 15-ft. Once white sapote trees become 30 ft (9.1 m) or taller, extreme caution should be used in pruning the trees. Climbing trees to prune them is dangerous and not recommended. Pruning of large white sapote trees should be done by a professional arborist that is licensed and insured.

Harvest, Ripening, and Storage

Mature fruits may be picked several days before natural fruit drop occurs and should be clipped, leaving a small piece of stem attached. When the fruit becomes ripe, the stem falls off. Fruit should be handled very carefully because the peel is very thin and susceptible to damage, and the pulp below bruises generally becomes bitter. Harvested fruit should be ripened at room temperature (26-28°C). Once ripe it may be stored for 7 to 10 days in the refrigerator.

Papaya

The papaya (from Carib via Spanish), papaw, or pawpaw is the fruit of the plant *Carica papaya*, the sole species in the genus *Carica* of the plant family Caricaceae. It is native to the tropics of the Americas, perhaps from southern Mexico and neighbouring Central America. It was first cultivated in Mexico several centuries before the emergence of the Mesoamerican classical civilizations.

The papaya is a large, tree-like plant, with a single stem growing from 5 to 10 m (16 to 33 ft) tall, with spirally arranged leaves confined to the top of the trunk. The lower trunk is conspicuously scarred where leaves and fruit were borne. The leaves are large, 50–70 cm (20–28 in) in diametre, deeply palmately lobed, with seven lobes. Unusually for such large plants, the trees are dioecious. The tree is usually unbranched, unless lopped. The flowers are similar in shape to the flowers of the *Plumeria*, but are much smaller and wax-like. They appear on the axils of the leaves, maturing into large fruit - 15–45 cm (5.9–17.7 in) long and 10–30 cm (3.9–11.8 in) in diametre. The fruit is ripe when it feels soft (as soft as a ripe avocado or a bit softer) and its skin has attained an amber to orange hue.

Carica papaya was the first transgenic fruit tree to have its genome deciphered.

Cultivation

Gaining in popularity among tropical fruits worldwide, papaya is now ranked third with 11.22 Mt, or 15.36 percent of the total tropical fruit production, behind mango with 38.6 Mt (52.86%) and pineapple with 19.41 Mt (26.58%). Global papaya production has grown significantly over the last few years, mainly as a result of increased production in India. Papaya has become an important agricultural export for developing countries, where export revenues of the fruit provide a livelihood for thousands of people, especially in Asia and Latin America. Papaya exports contribute to the growing supply of healthy food products on international markets. The top three exporting countries accounted for 63.28 percent of the total global exports of papaya between 2007 and 2009, with more than half of those exports going to the United States.

Global papaya production is highly concentrated, with the top ten countries averaging 86.32 percent of the total production for the period 2008–2010. India is the leading papaya producer, with a 38.61 percent share of the world production during 2008–2010, followed by Brazil (17.5%) and Indonesia (6.89%). Other important papaya producing countries and their share of global production include Nigeria (6.79%), Mexico (6.18%), Ethiopia (2.34%), Democratic Republic of the Congo (2.12%), Colombia (2.08%), Thailand (1.95%), and Guatemala (1.85%).

Originally from southern Mexico (particularly Chiapas and Veracruz), Central America, and northern South America, the papaya is now cultivated in most tropical countries. In cultivation, it grows rapidly, fruiting within three years. It is, however, highly frost-sensitive, limiting its production to tropical climates. Temperatures below –2 °C (29 °F) are greatly harmful if not fatal. In Florida, growth is generally limited to southern parts of the state. It also prefers sandy, well-drained soil as standing water will kill the plant within 24 hours.

For cultivation, however, only female plants are used, since they give off a single flower each time, and close to the base of the plant, while the male gives off multiple flowers in long stems, which result in poorer quality fruit.

Diseases and Pests

Viruses: Papaya ringspot virus is a well-known virus within plants in Florida. The first signs of the virus are yellowing and vein-clearing of younger leaves as well as mottling yellow leaves. Infected leaves may obtain blisters, roughen or narrow, with blades sticking upwards from the middle of the leaves. The petioles and stems may develop dark green greasy streaks and in time become shorter. The ringspots are circular, C-shaped markings that are darker green than the fruit itself. In the later stages of the virus, the markings may become gray and crusty. Viral infections impact growth and reduce the fruit's quality. One of the biggest effects that viral infections have on papaya is the taste. Right now, the virus is uncontrollable.

The papaya mosaic virus is very destructive because it completely destroys the plant until only a small tuft of leaves are left. The virus affects both the leaves of the plant and also the fruit itself. Leaves show thin, irregular, dark-green lines around the borders and clear areas around the veins. The more severely affected leaves are irregular and linear in shape. The virus can infect the fruit at any stage of its maturity. Fruits as young as 2 weeks old have been spotted with dark-green ringspots about 1 inch in diametre. Rings on the fruit are most likely seen on either the stem end or the blossom end. In the early stages of the ringspots, the rings tend to be many closed circles, but as the disease develops the rings will increase in diametre consisting of one large ring. The difference between the ringspot and the mosaic viruses is the ripe fruit in the ringspot has mottling of colours and mosaic doesn't.

Fungi: The fungus Anthracnose is known to specifically attack papaya especially the mature fruits. The disease starts out small with very few signs, such as water-soaked spots on ripening fruits. The spots become sunken, turn brown or black and may get bigger. In some of the older spots, the fungus may produce pink spores. The fruit ends up being soft and having an off flavor because the fungus grows into the fruit.

The fungus powdery mildew occurs as a superficial white presence on the surface of the leaf in which it is easily recognized. Tiny, light yellow spots begin on the lower surfaces of the leaf as the disease starts to make its way. The spots enlarge and white powdery growth appears on the leaves. The infection usually appears at the upper leaf surface as white fungal growth. Powdery Mildew is not as severe as other diseases.

The fungus phythphthora blight causes damping-off, root rot, stem rot, stem girdling and fruit rot. Damping-off happens in very young plants by wilting and death in plant. The spots on established plants start out as water-soaked lesions at the fruit and branch scars. These spots can get bigger and cause the death of the plant. The roots can be severely and rapidly infected, causing the plant to rapidly brown and wilt away collapsing within days. The most dangerous feature of the disease is the infection of the fruit because it cause harm to people who consume it. The biggest evidence that the fungus is present are the water-soaked marks that appears first along with the white fungus that grows on the dead fruit. After the fruit dies it shrivels and falls to the ground.

Pests: The papaya fruit fly is mainly yellow with black marks. The female papaya fruit fly has a very long, slender abdomen with an extended ovipositor that exceeds the length of its body. The male papaya fruit fly looks like the female with the differences of a hairy abdomen and no ovipositor. Long slender eggs are laid inside of the fruit by the female papaya fruit fly. The larva are white and look very much like the regular fruit fly larvae. The female is capable of laying up to 100 or more eggs and are laid during the evening or early morning in groups of ten inside young fruit. They usually hatch within 12 days of being in the fruit where they'll feed on the seeds and interior parts of the fruit. When the larvae matures (usually 16 days after being hatched) they eat their way out of the fruit, drop to the ground, and pupate just below the soil and emerge within one to two weeks as mature flies. The flesh of the papaya must be ripe in order for the fly to migrate towards the surface of the fruit because unripe papaya juice is fatal to them. The papaya will turn yellow and drop to the ground if it is infected by the papaya fruit fly.

The two-spotted spider mite is a 0.5 mm long brown or orange-red but a green, greenish yellow translucent oval pest. They all have needle-like piercing-sucking mouthparts and feed by piercing the plant tissue with their mouth parts usually the underside of the plant. The spider mites spin fine threads of webbing on the host plant and when they remove the sap, the mesophyll tissue collapses and a small chlorotic spot forms at the feeding sites. The leaves of the papaya fruit turn yellow, gray or bronze. If the spider mites aren't controlled it can cause the death of the fruit.

The papaya whitefly lays yellow oval eggs that appear dusted on the undersides of the leaves. They eat the papaya fruits leaves therefore damaging the fruit. There, the eggs developed into flies in three stages called instars. The first instar has well-developed legs and is the only mobile immature life stage. The crawlers insert their mouthparts in the lower surfaces of the leaf when they find it suitable and usually don't move again in this stage. The next instars are flattened, oval and scale-like. In the final stage as the pupal the whiteflies are more convex, with large conspicuous red eyes.

Cultivars: Two kinds of papayas are commonly grown. One has sweet, red or orange flesh, and the other has yellow flesh; in Australia, these are called "red papaya" and "yellow papaw", respectively. Either kind, picked green, is called a "green papaya."

The large-fruited, red-fleshed 'Maradol', 'Sunrise', and 'Caribbean Red' papayas often sold in US markets are commonly grown in Mexico and Belize.

In 2011 Philippine researchers reported that by hybridizing papaya with *Vasconcellea quercifolia* they had developed conventionally bred, nongenetically engineered papaya resistant to PRV.

Genetically Engineered Cultivars

In response to the papaya ringspot virus (PRV) outbreak in Hawaii, genetically altered papaya were generated and brought to market (including 'SunUp' and 'Rainbow') that have some PRV DNA incorporated into the DNA of the plant are resistant to PRVs. This was so successful that by 2010, 80% of Hawaiian papaya plants were genetically modified.

3

Sub-Tropical Fruits

Banana

A banana is an edible fruit produced by several kinds of large herbaceous flowering plants in the genus *Musa*. (In some countries, bananas used for cooking may be called plantains.) The fruit is variable in size, colour and firmness, but is usually elongated and curved, with soft flesh rich in starch covered with a rind which may be green, yellow, red, purple, or brown when ripe. The fruits grow in clusters hanging from the top of the plant. Almost all modern edible parthenocarpic (seedless) bananas come from two wild species – *Musa acuminata* and *Musa balbisiana*. The scientific names of most cultivated bananas are *Musa acuminata*, *Musa balbisiana*, and *Musa* × *paradisiaca* for the hybrid *Musa acuminata* × *M. balbisiana*, depending on their genomic constitution. The old scientific name *Musa sapientum* is no longer used.

Musa species are native to tropical Indomalaya and Australia, and are likely to have been first domesticated in Papua New Guinea. They are grown in at least 107 countries, primarily for their fruit, and to a lesser extent to make fibre, banana wine and banana beer and as ornamental plants.

Worldwide, there is no sharp distinction between "bananas" and "plantains". Especially in the Americas and Europe, "banana" usually refers to soft, sweet, dessert bananas, particularly those of the Cavendish group, which are the main exports from banana-growing countries. By contrast, *Musa* cultivars with firmer, starchier fruit are called "plantains". In other regions, such as Southeast Asia, many more kinds of banana are grown and eaten, so the simple two-fold distinction is not useful and is not made in local languages.

The term "banana" is also used as the common name for the plants which produce the fruit. This can extend to other members of the genus *Musa* like the scarlet banana (*Musa coccinea*), pink banana (*Musa velutina*) and the Fe'i bananas. It can also refer to members of the genus *Ensete*, like the snow banana (*Ensete glaucum*) and the economically important false banana (*Ensete ventricosum*). Both genera are classified under the banana family, Musaceae.

The banana plant is the largest herbaceous flowering plant. All the above-ground parts of a banana plant grow from a structure usually called a "corm". Plants are normally tall and fairly sturdy, and are often mistaken for trees, but what appears to be a trunk is actually a "false stem" or pseudostem. Bananas grow in a wide variety of soils, as long as the soil is at least 60 cm deep, has good drainage and is not compacted. The leaves of banana plants are composed of a "stalk" (petiole) and a blade (lamina). The base of the petiole widens to form a sheath; the tightly packed sheaths make up the pseudostem, which is all that supports the plant. The edges of the sheath meet when it is first produced, making it tubular. As new growth occurs in the centre of the pseudostem the edges are forced apart. Cultivated banana plants vary in height depending on the variety and growing conditions. Most are around 5 m (16 ft) tall, with a range from 'Dwarf Cavendish' plants at around 3 m (10 ft) to 'Gros Michel' at 7 m (23 ft) or more. Leaves are spirally arranged and may grow 2.7 metres (8.9 ft) long and 60 cm (2.0 ft) wide. They are easily torn by the wind, resulting in the familiar frond look.

When a banana plant is mature, the corm stops producing new leaves and begins to form a flower spike or inflorescence. A stem develops which grows up inside the pseudostem, carrying the immature inflorescence until eventually it emerges at the top. Each pseudostem normally produces a single inflorescence, also known as the "banana heart". (More are sometimes produced; an exceptional plant in the Philippines produced five.) After fruiting, the pseudostem dies, but offshoots will normally have developed from the base, so that the plant as a whole is perennial. In the plantation system of cultivation, only one of the offshoots will be allowed to develop in order to maintain spacing. The inflorescence contains many bracts (sometimes incorrectly referred to as petals) between rows of flowers. The female flowers (which can develop into fruit) appear in rows further up the stem (closer to the leaves) from the rows of male flowers. The ovary is inferior, meaning that the tiny petals and other flower parts appear at the tip of the ovary. The banana fruits develop from the banana heart, in a large hanging cluster, made up of tiers (called "hands"), with up to 20 fruit to a tier. The hanging cluster is known as a bunch, comprising 3–20 tiers, or commercially as a "banana stem", and can weigh from 30–50 kilograms (66–110 lb). Individual banana fruits (commonly known as a banana or "finger") average 125 grams (0.276 lb), of which approximately 75% is water and 25% dry matter.

The fruit has been described as a "leathery berry". There is a protective outer layer (a peel or skin) with numerous long, thin strings (the phloem bundles), which run lengthwise between the skin and the edible inner portion. The inner part of the common yellow dessert variety splits easily lengthwise into three sections that correspond to the inner portions of the three carpels. In cultivated varieties, the seeds are diminished nearly to non-existence; their remnants are tiny black specks in the interior of the fruit.

Bananas are naturally slightly radioactive, more so than most other fruits, because of their potassium content and the small amounts of the isotope potassium-40 found in naturally occurring potassium. The banana equivalent dose of radiation is sometimes used in nuclear communication to compare radiation levels and exposures.

Bananas and Plantains

In regions such as North America and Europe, *Musa* fruits offered for sale can be divided into "bananas" and "plantains", based on their intended use as food. Thus the banana producer and distributor Chiquita produces publicity material for the American market which says that "a plantain is not a banana". The stated differences are that plantains are more starchy and less sweet; they are eaten cooked rather than raw; they have thicker skin, which may be green, yellow or black; and they can be used at any stage of ripeness. Linnaeus made the same distinction between plantains and bananas when first naming two "species" of *Musa*. Members of the "Plantain subgroup" of banana cultivars, most important as food in West Africa and Latin America, correspond to the Chiquita description, having long pointed fruit. They are described by Ploetz et al. as "true" plantains, distinct from other cooking bananas. The cooking bananas of East Africa belong to a different group, the East African Highland bananas, so would not qualify as "true" plantains on this definition.

An alternative approach divides bananas into dessert bananas and cooking bananas, with plantains being one of the subgroups of cooking bananas. Triploid cultivars derived solely from *M. acuminata* are examples of "dessert bananas", whereas triploid cultivars derived from the hybrid between *M. acuminata* and *M. balbinosa* (in particular the Plantain subgroup of the AAB Group) are "plantains". Small farmers in Colombia grow a much wider range of cultivars than large commercial plantations. A study of these cultivars showed that they could be placed into at least three groups based on their characteristics: dessert bananas, non-plantain cooking bananas, and plantains, although there were overlaps between dessert and cooking bananas.

In Southeast Asia – the centre of diversity for bananas, both wild and cultivated – the distinction between "bananas" and "plantains" does not work, according to Valmayor et al. Many bananas are used both raw and cooked. There are starchy cooking bananas which are smaller than those eaten raw. The range of colours, sizes and shapes is far wider than in those grown or sold in Africa, Europe or the Americas. Southeast Asian languages do not make the distinction between "bananas" and "plantains" that is made in English (and Spanish). Thus both Cavendish cultivars, the classic yellow dessert bananas, and Saba cultivars, used mainly for cooking, are called *pisang* in Malaysia and Indonesia, *kluai* in Thailand and *chuoi* in Vietnam. Fe'i bananas, grown and eaten in the islands of the Pacific, are derived from entirely different wild species than traditional bananas and plantains. Most Fe'i bananas are cooked, but Karat bananas, which are short and squat with bright red skins, very different from the usual yellow dessert bananas, are eaten raw.

In summary, in commerce in Europe and the Americas (although not in small-scale cultivation), it is possible to distinguish between "bananas", which are eaten raw, and "plantains", which are cooked. In other regions of the world, particularly India, Southeast Asia and the islands of the Pacific, there are many more kinds of banana and the two-fold distinction is not useful and not made in local languages. Plantains are one of many kinds of cooking bananas, which are not always distinct from dessert bananas.

Modern Cultivation

All widely cultivated bananas today descend from the two wild bananas *Musa acuminata* and *Musa balbisiana*. While the original wild bananas contained large seeds, diploid or polyploid cultivars (some being hybrids) with tiny seeds are preferred for human raw fruit consumption. These are propagated asexually from offshoots. The plant is allowed to produce two shoots at a time; a larger one for immediate fruiting and a smaller "sucker" or "follower" to produce fruit in 6–8 months. The life of a banana plantation is 25 years or longer, during which time the individual stools or planting sites may move slightly from their original positions as lateral rhizome formation dictates.

Cultivated bananas are parthenocarpic, i.e. the flesh of the fruit swells and ripens without its seeds being fertilized and developing. Lacking viable seeds, propagation typically involves farmers removing and transplanting part of the underground stem (called a corm). Usually this is done by carefully removing a sucker (a vertical shoot that develops from the base of the banana pseudostem) with some roots intact. However, small sympodial corms, representing not yet elongated suckers, are easier to transplant and can be left out of the ground for up to two weeks; they require minimal care and can be shipped in bulk.

It is not necessary to include the corm or root structure to propagate bananas; severed suckers without root material can be propagated in damp sand, although this takes somewhat longer.

In some countries, commercial propagation occurs by means of tissue culture. This method is preferred since it ensures disease-free planting material. When using vegetative parts such as suckers for propagation, there is a risk of transmitting diseases (especially the devastating Panama disease).

As a non-seasonal crop, bananas are available fresh year-round.

Pests, Diseases and Natural Disasters

Figure: *Banana bunches are sometimes encased in plastic bags for protection. The bags may be coated with pesticides.*

While in no danger of outright extinction, the most common edible banana cultivar Cavendish (extremely popular in Europe and the Americas) could become unviable for large-scale cultivation in the next 10–20 years. Its predecessor 'Gros Michel', discovered in the 1820s, suffered this fate. Like almost all bananas, Cavendish lacks genetic diversity, which makes it vulnerable to diseases, threatening both commercial cultivation and small-scale subsistence farming. Some commentators remarked that those variants which could replace what much of the world considers a "typical banana" are so different that most people would not consider them the same fruit, and blame the decline of the banana on monogenetic cultivation driven by short-term commercial motives.

Panama Disease

Panama disease is caused by a fusarium soil fungus (Race 1), which enters the plants through the roots and travels with water into the trunk and leaves, producing gels and gums that cut off the flow of water and nutrients, causing the plant to wilt, and exposing the rest of the plant to lethal amounts of sunlight. Prior to 1960, almost all commercial banana production centred on "Gros Michel", which was highly susceptible. Cavendish was chosen as the replacement for Gros Michel because, among resistant cultivars, it produces the highest quality fruit. However, more care is required for shipping the Cavendish, and its quality compared to Gros Michel is debated.

According to current sources, a deadly form of Panama disease is infecting Cavendish. All plants are genetically identical, which prevents evolution of disease resistance. Researchers are examining hundreds of wild varieties for resistance.

Tropical Race 4

Tropical race 4 (TR4) is a reinvigorated strain of Panama disease first discovered in 1993. This virulent form of fusarium wilt has wiped out Cavendish in several southeast Asian countries. It has yet to reach the Americas; however, soil fungi can easily be carried on boots, clothing, or tools. This is how TR4 travels and is its most likely route into Latin America. Cavendish is highly susceptible to TR4, and over time, Cavendish is almost certain to be eliminated from commercial production by this disease. The only known defence to TR4 is genetic resistance.

Black Sigatoka

Black sigatoka is a fungal leaf spot disease first observed in Fiji in 1963 or 1964. Black Sigatoka (also known as black leaf streak) has spread to banana plantations throughout the tropics from infected banana leaves that were used as packing material. It affects all main cultivars of bananas and plantains (including the Cavendish cultivars), impeding photosynthesis by blackening parts of the leaves, eventually killing the entire leaf. Starved for energy, fruit production falls by 50% or more, and the bananas that do grow ripen prematurely, making them unsuitable for export. The fungus has shown ever-increasing resistance to treatment, with the current expense for treating 1 hectare (2.5 acres) exceeding $1,000 per year. In addition to the expense, there is the question of how long intensive spraying can be environmentally justified. Several resistant cultivars of banana have been developed, but none has yet received commercial acceptance due to taste and texture issues.

In East Africa

With the arrival of Black sigatoka, banana production in eastern Africa fell by over 40%. For example, during the 1970s, Uganda produced 15 to 20 tonnes (15 to 20 long tons; 17 to 22 short tons) of bananas per hectare. Today, production has fallen to only 6 tonnes (5.9 long tons; 6.6 short tons) per hectare.

The situation has started to improve as new disease-resistant cultivars have been developed by the International Institute of Tropical Agriculture and the National Agricultural Research Organisation of Uganda (NARO), such as FHIA-17 (known in Uganda as the Kabana 3). These new cultivars taste different from the Cabana banana, which has slowed their acceptance by local farmers. However, by adding mulch and manure to the soil around the base of the plant, these new cultivars have substantially increased yields in the areas where they have been tried.

The International Institute of Tropical Agriculture and NARO, funded by the Rockefeller Foundation and CGIAR have started trials for genetically modified bananas that are resistant to both Black sigatoka and banana weevils. It is developing cultivars specifically for smallholder and subsistence farmers.

Banana Bunchy Top Virus

Banana bunchy top virus (BBTV) jumps from plant to plant using aphids. It stunts leaves, resulting in a "bunched" appearance. Generally, an infected plant does not produce fruit, although mild strains exist which allow some production. These mild strains are often mistaken for malnourishment, or a disease other than BBTV. There is no cure; however, its effect can be minimized by planting only tissue-cultured plants (in vitro propagation), controlling aphids, and immediately removing and destroying infected plants.

Banana Bacterial Wilt

Banana bacterial wilt (BBW) is a bacterial disease caused by *Xanthomonas campestris* pv. *musacearum*. After being originally identified on a close relative of bananas, *Ensete ventricosum*, in Ethiopia in the 1960s, BBW occurred in Uganda in 2001 affecting all banana cultivars. Since then BBW has been diagnosed in Central and East Africa including the banana growing regions of Rwanda, the Democratic Republic of the Congo, Tanzania, Kenya, Burundi, and Uganda.

Susceptibility

The fungus causes significant damage to banana crops due to its virulence, as well as the identical genetic composition of modern banana crops. Modern breeds of bananas cannot reproduce sexually because they have no seeds and the male flowers do not produce viable pollen. New banana plants develop through asexual reproduction: after the fruiting stem has matured, fruited, and been cut down, the base of the plant produces suckers, which can be cut off and planted elsewhere. This is like taking a cutting from a rose bush; the cuttings have to be moved and planted by humans.

As a result, these banana plants are nearly identical genetically, thus have the same susceptibility to the same diseases. Once a pathogen overcomes the plant's defenses, it can quickly infect the whole cultivated area. That happened to the Gros Michel banana. Over several decades, the fungus spread from Panama to neighbouring countries, moving north through Costa Rica to Guatemala and south into Colombia and Ecuador. By 1960, 50 years after the disease was discovered, the Gros Michel banana was, in effect, wiped out.

The Passion Fruit

Passiflora edulis is a vine species of passion flower that is native to Brazil, Paraguay, Uruguay and northern Argentina (Corrientes and Misiones provinces, among others). Its common names include passion fruit (US), passionfruit (UK and Commonwealth), and purple granadilla (South Africa).

It is cultivated commercially in warmer, frost-free areas for its fruit and is widely grown in Antigua, Argentina, Australia, Bolivia, Brazil, the Caribbean, Colombia, Costa Rica, Dominican Republic, East Africa, Ecuador, Haiti, Honduras, India, Indonesia, Israel, Mexico, New Zealand, Panama, Peru, Portugal (Madeira), Puerto Rico, Sri Lanka, South Africa, United States (California, Florida, Hawaii), Venezuela, Philippines, Malaysia.

The passion fruit is round to oval, either yellow or dark purple at maturity, with a soft to firm, juicy interior filled with numerous seeds. The fruit is both eaten and juiced; passion fruit juice is often added to other fruit juices to enhance the aroma.

Varieties

Several distinct varieties of passion fruit with clearly differing exterior appearances exist. The bright yellow *flavicarpa* variety, also known as the Golden Passion Fruit, can grow up to the size of a grapefruit, has a smooth, glossy, light and airy rind, and has been used as a rootstock for the Purple Passion Fruit in Australia. The dark purple *edulis* variety is smaller than a lemon, though it is less acidic than the yellow passion fruit, and has a richer aroma and flavour.

The purple varieties of the fruit have been found to contain traces of cyanogenic glycosides in the skin.

Uses

- In Australia and New Zealand, where it is called "passionfruit", it is available commercially both fresh and tinned. It is added to fruit salads, and fresh fruit pulp or passion fruit sauce is commonly used in desserts, including as a topping for pavlova (a regional meringue cake) and ice cream, a flavouring for cheesecake, and in the icing of vanilla slices. A passionfruit-flavoured soft drink called Passiona has also been manufactured in Australia since the 1920s.
- In Brazil, the term *maracujá* applies to passion fruit (*maracujá azedo*, or "sour") and granadillo (*maracujá doce*, or "sweet"). Passion fruit mousse is a common dessert, and passion fruit pulp is routinely used to decorate the tops of cakes. Passion fruit juice, ice pops and more recently soft drinks are also very popular. When making caipirinha, it is usual to use passion fruit instead of lime; it is then called *maracujá*. It is used also as a mild sedative, and its active ingredient is commercialized under several brands, most notably Maracugina.
- In Colombia it is one of the most important fruits, especially for juices and desserts. It is widely available all over the country and three kinds of "Maracuyá" fruit may be found.
- In the Dominican Republic, where it is locally called *chinola*, it is used to make juice and Fruit preserves. Passion fruit-flavoured syrup is used on shaved ice, and the fruit is also eaten raw, sprinkled with sugar.

- In Indonesia there are two types of passionfruit (local name: 'markisa'), white flesh and yellow flesh. The white one is normally eaten straight as a fruit, while the yellow variety is commonly strained to obtain its juice, which is cooked with sugar to make thick syrup. Bottles or plastic jugs of concentrated syrup (generally produced in Sumatra from fruit grown in the Lake Toba region) are sold in many supermarkets. Dilution of one part syrup to four (or more) parts water is recommended.
- In Israel passion fruit is used to make fruit wine.
- In Mexico passion fruit is used to make juice or is eaten raw with chilli powder and lime.
- In Paraguay passion fruit is used principally for its juice, to prepare desserts such as passion fruit mousse, cheesecake, ice cream, and to flavour yogurts and cocktails.
- In Peru passion fruit is used in several desserts, especially cheesecakes. Passion fruit juice is also drunk on its own and is used in ceviche variations and in cocktails, including the *Maracuyá Sour*, a variation of the Pisco Sour.
- In the Philippines passion fruit is commonly sold in public markets and in public schools. Some vendors sell the fruit with a straw to enable sucking out the seeds and juices inside. It is not very popular because of its sour flavour, and the fruit is very seasonal.
- In Portugal, especially the Azores and Madeira, passion fruit is used as a base for a variety of liqueurs and mousses.
- In Puerto Rico, where the fruit is known as "Parcha", it is widely believed to lower blood pressure, probably because it contains harmala alkaloids and is a mild RIMA. Passion fruit juice is also very common there and is used in juices, ice cream or pastries.
- In South Africa passion fruit, known locally as Granadilla (the yellow variety as Guavadilla), is used to flavour yogurt. It is also used to flavour soft drinks such as Schweppes' "Sparkling Granadilla" and numerous cordial drinks. It is often eaten raw or used as a topping for cakes and tarts. Granadilla juice is commonly available in restaurants. The yellow variety is used for juice processing, while the purple variety is sold in fresh-fruit markets.
- In Sri Lanka passion fruit juice, along with faluda, is one of the most popular refreshments. Passion fruit cordial is manufactured both at home as well as industrially by mixing the pulp with

sugar. There are many cordial manufacturers, suppliers and exporters in the country.

- In Thailand passion fruit is called "Saowarot". The fruit is eaten whole and is also commonly juiced and drunk. Young shoots are cooked in curries or eaten with nam phrik.
- In the United States it is often used as an ingredient in alcoholic beverages and juice mixes.
  - o In Hawaii passion fruit is called *lilikoi* and comes in yellow and purple varieties.

Passion fruit can be cut in half and the seeds scooped out with a spoon. Lilikoi-flavoured syrup is a popular topping for shave ice. It is used as a dessert flavouring for malasadas, cheesecakes, cookies, ice cream and mochi. Passion fruit is also favoured as a jam or jelly, as well as a butter. Lilikoi syrup can also be used to glaze or marinade meat and vegetables. Most passion fruit comes from backyard gardens or is collected from the wild. While it may be found at farmers' markets throughout the islands, fruits are seldom sold in grocery stores.

- In Vietnam passion fruit is blended with honey and ice to create refreshing smoothies.
- In Cambodia passion fruit is called "Machu Bey-darch", and the plant vine grows in the wild. Bushes hang with green to yellow round fruits, measuring from 2.5 cm to 4 cm when ripe. This wild variety of passion fruit tastes slightly different but is still quite sour.
- In India the government of Andhra Pradesh started growing passion fruits in the Chintapalli (Vizag) region forests to make them available to the local people. However, the fruit is found in the jungles of Assam and is known to local people as "Lota Bel."
- In Costa Rica it is known as "Estococa". The fruit grows in the wild, and it is commonly used for juice, it is considerably smaller than the Maracuyá.

Nutrition

Fresh passion fruit contains provitamin A beta carotene, vitamin C (36%), dietary fibre (42%) and iron (12%) in significant quantities as percent of the Daily Value; the vitamin A content converted from provitamin A sources is 25%. Passion fruit juice is a good source of potassium, possibly making the fruit relevant as a nutrient source for lowering risk of high blood pressure. Preliminary research indicated that consuming passion fruit peel may relieve asthma symptoms. One report showed that the fruit pericarp contains lycopene.

Climatic Conditions, Soil and Water Management

The yellow passion fruit grows best at altitudes of 0-800 m and is used mainly for the fresh fruit market. It can also be used as a rootstock for grafting of the purple variety; the purple passion fruit forms virtually no flowers below 1000 m and should be grown at altitudes of 1200-2000m.

The mature purple passion fruit tolerates light frosts and can be grown in the subtropics. Other varieties exist such as the so-called banana passion fruit growing in highland areas and often climb very tall trees. It is yellow at maturity but with soft velvety skin and pink flowers quite different from the commercial passion fruits. The banana passion fruit is mainly used in sweetened juices, as it is not usually very sweet on its own. All varieties grow on a wide range of soils; but light to heavy sandy loams, of medium texture and at least 60 cm deep are most suitable. Heavy clay soils have to be drained and very sandy ones need heavy manuring. A pH of 5.5-7 is preferred. If the soil is too acidic, lime must be applied. Good drainage and aeration are essential to minimise the incidence of diseases such as collar rot.

Purple passion fruit, especially, grows well on as little as 900 mm rainfall in Africa, provided the rainfall is well distributed. The vines require sheltered locations without extreme temperatures: Optimum temperatures for the purple variety are between 18-25°C and for the yellow variety 25-30°C. Critical temperatures were established for hybrid cultivars in Australia as follows: below 20°C pollen does not germinate and at 18-15°C both growth and flowering are set back, whereas temperatures above 30-32°C stimulate growth at the expense of flowering and fruit set (CABI).

Passion Fruit Types

There are two main forms of the passion fruit, the purple and the yellow varieties, with the yellow being distinguished as P. edulis var.flavicarpa. Hybrids can be made between purple and yellow passion fruit, which yields intermediates between the two forms.

Principal characteristics of the yellow type:

- Yellow rind and larger fruit
- More acid flavour
- Flowers are self-sterile - wind is ineffective because of the heaviness and stickiness of the pollen. They must be pollinated, and carpenter bees are the most efficient pollinators.

- More vigorous vine
- More tolerant of frost
- Resistant to nematodes and Fusarium wilt
- Brown seeds

Principal characteristics of the purple type:

- Purple rind and smaller fruit
- Sweet less acidic pulp richer in aroma and flavour and has a higher proportion of juice (35-38%)
- Can self pollinate but pollination is best under humid conditions
- Less vigorous vine
- If crossing yellow and purple types, it is necessary to use the purple parent as the seed parent because the flowers of the yellow are not receptive to the pollen of the purple, and an early blooming yellow must be utilised in order to have a sufficient overlapping period for pollen transfer.
- Black seeds

Propagation and Planting

Passion fruit is generally propagated from seed, although cuttings and grafting can be used. Seed should be rubbed clean of pulp and dried in the shade. Germination takes 2-4 weeks. Fresh seeds are much easier to germinate than seeds older than one or two months. Older seeds can be soaked for at least one day to improve germination. Seedlings are often raised in polythene bags, 15 cm wide and 25 cm deep. Three seeds per bag are sown at a depth of 1 cm and thinned to leave one after two months. Cuttings are set in coarse sand and later transplanted into bags or a nursery bed. The seedlings grow slowly and require 3-4 months to reach the transplanting height of 15-25 cm. Seedlings must be hardened off by leaving them in an open, shaded area for a day or two.

Grafting is often used to control diseases. Yellow passion fruit is used as resistant rootstock although other Passiflora species, in particular P. caerulea L., show much greater resistance to Phytophthora root rot and Fusarium collar rot. Moreover, P. caerulea is tolerant of root-knot nematodes and to exposure to -1.5°C; it can be propagated from leaf and stem cuttings and is compatible with P. edulis. Wedge and whip grafts on seedling rootstocks - sometimes on rooted cuttings - are used.

Within 5-7 weeks after transplanting, each plant will have up to four healthy laterals. From then on the vine grows very rapidly; the first flowers are produced 5-7 months after transplanting when the vine can be 10-15 m long.

Light is the essential factor for flowering and in passionfruit this is particularly true for floral development and fruit set. That is why training and pruning are important to ensure adequate exposure of the shoots. Depending on the climate there may be one to three harvest peaks (purple passionfruit) or a single, often very long harvest season (more common with the yellow passionfruit).

Land Preparation

Deep ploughing and harrowing is necessary to remove hard pans in the soil. Passion fruit has a deep root system; therefore proper land cultivation is necessary. Commercial plantations adopt a row spacing of 1.2-1.8 m and a within-row spacing of 3 m. This gives around 1900-2700 plants/ha. Planting holes of 45 x 45 x 45 cm should be filled with topsoil mixed with up to 10 kg of compost or manure. Transplanting is done at the start of the rainy season. At planting the soil around the plants should be firmed down to establish good root/soil contact. In order to avoid fungal infection the grafting spot should not have any contact with the soil during and after planting. The seedling should then be irrigated to ensure quick rooting and establishment of the plant.

Husbandry

Early growth of passion fruit is slow and regular weeding is essential. Care should be taken when weeding in order to avoid any injury to the plant. Mulching along the rows or around the base of the plants greatly facilitates weed control and protects the roots. Elaborate trellises have been used in Australia and South Africa, but in East Africa, especially at closer spacing, a single wire trellis has been found to be as good. A 14-gauge galvanized wire is tightly stretched along the tops of hardwood posts 15 cm in diametre and 3 m long, dug in to a depth of 0.6 m; these posts are spaced 8 m apart. The trellis should be erected when the field is planted so that the main shoot and one vigorous lateral can be tied to the wire with a string. If laterals do not emerge in time, they can be forced to leaf out by pinching off the shoot tip. When the vines reach the wire they are trained in opposite directions along it. All laterals below the wire are pruned off. Laterals emerging along the wire are allowed to hang down freely; they are the secondary shoots branching into tertiary shoots. Secondary and higher-order shoots are the fruiting wood, which has to be thinned and rejuvenated by pruning.

Regular fertilisation is necessary for optimum yields. Frequent sprays with compost tea or similar organic foliar feed should be applied starting from 1 month after planting and at least every 3 months after that. Mixing EM or BM with foliar sprays may prevent fungal attacks.

Pruning

Old unproductive shoots and dead wood must be removed. Also secondary shoots reaching the ground must be cut off about 5 cm above the ground. The laterals which bear fruit should be left to hang down freely from the wire and the entangling tendrils removed to allow free air and light penetration and reduce incidences of disease and pest epidemics. Disinfect with commercial detergent all equipment used for pruning regularly to avoid spread of viral diseases.

Intercropping

A wide range of vegetables and other crops can be intercropped with passion fruit. Intercropping with annuals is recommended; especially vegetables like beans, cabbages and tomatoes are agronomically suitable. Other recommended crops include potatoes, beetroots, Swiss chard, carrots, spinach, strawberries, eggplants, peppers, onions, leeks and head lettuce. However, cucurbits (cucumbers, pumpkin, and squashes) are not recommended due to the woodiness virus and fruit flies. Other crops that should not be intercropped with passion fruits are maize, cowpea, sorghum, okra, sweet potatoes and other creepers. Intercropping can help in erosion control particularly when fed with good compost.

Rotation

To avoid build up of soil-borne diseases strict crop rotation should be practised. Passion fruits should not be grown for more than 2-3 years on the same plot.

Yield

If a plantation is cropped for 3 years; of the total crop, roughly 50% is produced in the first year, 35% in the second, and 15% in the third year. The sharp decline in yield level, which is even more marked in areas with disease problems, is the main reason to replant fields after the second or third crop. Average yields amount to 10-15 t/ha per year for the purple and 20-25 t/ha per year for the yellow passionfruit. Much higher yields are possible; yields as high as 50 t/ha per year for purple passionfruit have been reported from Kenya.

Harvesting

Fruit drops to the ground when fully mature. It is collected every second day; at this stage it looks shrivelled and unattractive, but for processing fruits should be picked at this stage. For fresh fruit markets, especially the export market, fruit is picked after full colour development when the whole fruit is purple or canary yellow, but before shrivelling and drying set in.

Information on Pests

Biological methods of plant protection: Nematodes, especially the root-knot nematodes (Meloidogyne incognita, M. javanica and M. arenaria), are the most serious pests on passion fruit. Practical control measures are crop rotation and the use of tolerant rootstocks.

Several species of sucking bugs feed on passion fruit. They suck and pierce leaves and young fruits; these are minor pests.

Fruit flies that feed on passionfruit include the melon fly (B. cucurbitae) and the Mediterranean fruit fly (Ceratitis capitata) and the Queensland fruit fly (B. tryoni). Pierced young fruit shrivels and falls; later injuries cause damage which lowers the grade. Spraying of biopesticides may be necessary if destruction of infested fruit and the use of baits do not adequately check the pest.

Mealybugs (Planococcus citri and P. kenyae) are usually controlled by their natural enemies.

The same to applies mites the red spider mites and the broad mites - which incidentally do much damage.

Examples of Passion Fruit Pests and Organic Control Methods

Mealybugs (Planococcus citri and P. kenya) Mealybugs infest fruits and foliage. They can be serious pests in the warm season, if natural enemies, which usually control them, are destroyed by spraying with pesticides. They are 3-5 mm long, soft, elongate oval and somewhat flattened

What to do:

- Conserve natural enemies. Mealybugs are usually controlled by a wide range of natural enemies. However, use of pesticides may kill these natural enemies leading to mealybug outbreaks.

Aphids (Aphis gossypii and Myzus persicae) Aphids damage plants by sucking plant sap causing curling, wrinkling or cupping of infested leaves, giving plants a deformed shape. They spread viruses and excrete honeydew, which coat the plants and leads to growth of sooty mould, which can diminish the photosynthetic capacity of plants. Aphids are usually controlled by natural enemies if they have not been disturbed for instance by the use of broad-spectrum pesticides.

What to do:

- Plant the crop in well prepared, fertile land, but do avoid applying nitrogenous fertiliser, as this will promote new growth, which makes the plants juicy and attractive to aphids.

Leafmining flies (Lyriomyza spp) Feeding and egg laying by leafmining flies cause stippling of leaves. This can kill seedlings and in older plants allows entry of disease-causing microorganisms. Feeding by maggots causes mining (tunnelling) of leaves reducing the productive leaf area. Heavily attacked leaves may drop off, and may lead to yield losses.

What to do:

- Control by natural enemies is important.
- Ploughing can help in exposing pupae to desiccation and natural enemies.
- Neem products are effective for controlling leafminers.

Bugs Several species of sucking bugs feed on passion fruit. The most important are: The green stinkbug (Nezara viridula) The brown stinkbug (Boerias maculata)Coreid bugs such as the giant coreid bug or tip wilter (Anoplocnemis curvipes) and the leaf footed plant bug (Leptoglossus membranaceus).Bugs suck sap from the growing tips or developing fruits. The bugs pierce the terminal buds, which eventually wilt and die back. Young plants may be killed if the attack is severe. The punctured young fruits develop localised hardened spots that remain on the fruit reducing their market value.

What to do:

- In small orchard bugs can be hand picked and destroyed.
- Watering and irrigation discourage bugs.
- Old crops or sprouting stumps left in the field provide refuges for bugs so they should be destroyed or dig into the soil.
- Growing strong smelling plants such as garlic and onion near the crop is reported to reduce infestations.
- Spraying plants with a soapy solution helps to wash off young bugs.

Fruit flies (Bactrocera cucurbitae and Ceratitis capitata) Fruit flies that feed on passion fruit in Africa include the melon fly (Bactrocera cucurbitae) and the Mediterranean fruit fly (Ceratitis capitata). Pierced young fruit shrivels and falls; later injuries cause damage that lowers the market value of the fruit. However, the incidence of fruit flies on passion fruit is low, and usually of not economic importance. So control may not be necessary.

What to do:

- Collect and destroy all fallen fruits at least twice a week during the fruit season.
- Do not put collected damaged fruits into compost heaps. Instead, burn them or bury them at least 50 cm deep, so that the fruit flies cannot reach the soil surface.
- Remove fruits with dimples and those that ooze clear sap. This method is more labourious than picking the rotten fruits from the ground, but it is also more effective.
- Whenever possible, wrap fruit in newspaper or paper bags to prevent fruit flies from laying eggs on the fruit. This has to be done well before the fruit matures.
- Pick overripe fruits, as they attract fruit flies.
- Physical methods include fruit fly traps and fruit bagging.

Spider mites (Tetranychus spp.) Their feeding causes tiny yellow or white speckles, eventually leaves become yellowish and may drop, and may led to complete defoliation. Heavily infested plants may become stunted. A heavy infestation might also cause vine dieback and shrivelling and dropping of immature fruit.

What to do:

- Field hygiene is important for the management of spider mites. Old crops or weeds infested with mites can cause infestation of new crop.
- Natural enemies such as predatory mites are important for control of spider mites.

The broad mite or yellow tea mite (Polyphagotarsonemus latus) It is the most important mite pest of passion fruit in Kenya. Broad mites are tiny (0.1-0.2 mm long) and cannot be seen with the naked eye, and are even difficult to detect with a hand lens. An attack by the broad mites can be detected by the symptoms of damage. Their feeding produces discolouration, necrosis of tissues and deformation. Initial attack occurs on stems of terminal shoots and young terminal leaves. Attacked young leaves are stunted, deformed (slender, twisted or crumpled), fail to elongate and finally may wilt and dry. Stems of terminal shoots may become slightly swollen, roughened or russeted. As a result the growth of the plant is affected and flower production reduced causing considerably yield reduction. A bronzed dusty appearance may occur on affected plant parts. Attacked fruits become deformed and show white to tan or brown scars on the skin. This damage usually does not affect the internal quality of fruits but affect their market value. Severely attacked fruits may fall. Symptoms remain for a long period of time after control.

What to do:

- Broad mites are attacked by predacious mites. Phytoseiulus persimilis is not very much attracted to broad mites. Amblyseius spp. are better predators of broad mites, in particular A. californicus is used for control of broad mites in different parts of the world.
- Broad mites can be effectively controlled with sulphur sprays. However, sulphur is toxic to predatory mites and can have phytotoxic effects on young leaves and shoots at high sulphur dosages and when applied during hot weather.

Nematodes Amongst nematodes infesting passion fruit the root-knot nematodes (Meloidogyne incognita, M. javanica and M. arenaria), are the most serious pests. Characteristic symptoms of infestation by root-knot nematodes are formation of galls or knots on roots, yellowing of leaves, stunting and eventual wilting of the affected plants.

What to do:

- Rotate with cassava, cereals, maize, Baby corn, sweet corn, sweet potato, onions, cabbages / kale, garlic or fodder grasses (e.g. Sudan grass).
- Use of tolerant rootstocks (e.g. P. caerula).
- Maintain high organic matter (farmyard manure / compost) in the soil.
- Incorporation of neem into planting at transplanting.

Biological Methods of Plant Protection

The most important disease on passion fruit is brown spot (Alternaria passiflorae) on leaves, vines and fruits. Phytophthora blight (Phytophthora nicotianae) causes the wilting of shoot tips and crown rot, particularly where water stagnates occasionally. Septoria spot, caused by the fungus Septoria passiflorae, causes extensive spotting of leaf and fruit, and occasionally of the stem. Yellow passionfruit and its hybrids are more tolerant to these diseases.

Fusarium wilt (also called collar rot) is caused by the soil-borne fungus Fusarium oxysporum f.sp. passiflorae; the shoots wilt, followed by a complete collapse of the plant. Grafting to wilt-resistant yellow passionfruit rootstocks is the most practical way of control. Damping-off caused by Rhizoctonia solani and Pythium spp. can be a problem in nurseries and soils should be sterilised.

A number of virus diseases have been reported, notably passionfruit woodiness potyvirus (PWV). They are spread by aphids (Aphis gossypii and Myzus persicae) and pruning knives. Other virus diseases are ringspot from Côte d'Ivoire, which is similar to PWV. The most practical control is to use clean planting material, clean pruning tools and resistant hybrids, or rootstocks of yellow passionfruit.

Examples of Passion Fruit Diseases and Organic Control Methods

Brown Spot (Alternaria Passiflorae)

The most important disease worldwide is brown spot on leaves, vines and fruits. Symptoms are brown spots, up to 10 mm diametre, on the leaves, often extending along the veins and drying out in the centre. On the stems, spots are up to 30 mm long, and when they occur at the leaf axils may kill the vine, resulting in dieback. On the fruit, the spots are light brown, round and sunken; they often merge, covering large areas, and produce red-brown spore masses. Spores, produced on the leaf, stem and fruit, are dispersed by wind-blown rain. Warm, moist weather favours disease development.

What to do:

- Yellow passion fruit and its hybrids are more tolerant to this disease.
- Field sanitation (collection and disposal of fallen diseased fruits, leaves and vines).
- Pruning vines to reduce density and thereby reducing humidity within the crop. It also facilitate better air circulation, light and spray penetration and cover.
- Timely sprays with copper based fungicides. During humid weather, when the vines are growing rapidly, reduce the intervals between spray applications to 2 or 3 weeks to ensure that new growth is adequately protected.

Septoria Spot (Septoria passiflorae)

The disease attacks leaves, stems and fruits. Brown spots up to 2 mm with minute, black dots (fruiting bodies containing fungal spores) develop on leaf surface. Infected leaves fall readily leading to defoliation of vines. Similar spots may form on the stems albeit elongated. On fruits light-brown spots studded with minute black dots may be formed. The spots often join up to cover large areas of the fruit. Affected fruits ripen unevenly. Spores produced by black dots (fruiting bodies) are blown to adjacent vines during wet, windy weather thus further spreading the disease. The disease is spread by rain, dew and overhead irrigation. Warm moist weather favours disease development.

What to do: Disease management measures for brown spot disease are equally applicable for Septoria spot.

Fusarium Wilt (Fusarium Oxysporum f.sp. Passiflorae)

Fusarium wilt (also called collar rot) symptoms consist of yellowing of leaves, the collar region of affected plant at soil level turns brownish and vertically cracks and vines wilt followed by a complete collapse of the plant. On dissection of infected stem, vascular tissues show brown discolouration.

What to do:

- Affected parts should be removed and burned. Snap off the affected parts or remove the affected plant manually.
- Do not cut tissue and then use the knife on healthy plants.
- Keep the base of the plant clear of grass and weeds, which favour fungal growth.

Grafting to wilt-resistant yellow passion fruit rootstocks (e.g. P. caerula) is the most practical way of control.

Phytophthora Blight (Phytophthora Nicotianae var. Parastica)

Affected leaves are water-soaked and light-brown in colour. They fall readily, leading to defoliation of the vines. Affected areas of the stem are first purple and later brown above the graft union. They may completely girdle the stem causing wilting and collapse of the vine. Fruit symptoms comprise of large, water-soaked areas. Diseased fruits fall readily and in wet weather become covered with white, fungal growth.

Another strain of the fungus (Phytophthora cinnamoni) causes root rot. Yellow and purple varieties have different patterns of susceptibility. The yellow vine is susceptible to P. cinnamoni, and the purple vine is more susceptible to P. nicotianae. Both fungus strains attack both passion fruits and can cause root rot, wilt, damping off and leaf blight. Fungal spores are initially produced in wet soil beneath the vines and are splashed up to lower leaf canopy.

The disease is favoured by wet, windy weather.

What to do:

- Good field sanitation.

- Pruning and keeping a grass sward under the vines to minimise spore splashed up to the lower leaves.
- Graft to resistant rootstocks (e.g.)P. caerula.

The application of copper-based fungicides every 2-3 months during the wet season reduces disease incidence in areas where the disease is likely to be serious. Stem lesions may be painted with a copper fungicide.

Passion Fruit Woodiness Potyvirus (PWV)

A number of virus diseases have been reported, notably the passion fruit woodiness potyvirus. Affected leaves show light and dark green mosaic pattern often with light yellow speckle. Sometimes small, yellow ring spots may develop on upper leaf surface. Infected fruits are small and misshapen with very hard rind and small pulp cavity. When affected fruit is cut, the inside rind tissue may have brown spots. Some strains of the virus cause cracking of affected fruits.

They are spread by aphids (Aphis gossypii and Myzus persicae) grafting and pruning knives. The virus has a wide host range including bananas, cucurbits and many weeds.

What to do:

- Use virus-free planting material.
- Disinfect pruning tools with household bleach.
- Use resistant hybrids, or rootstocks of yellow passion fruit.
- Remove diseased vines from the field.
- Do proper weeding.

Avoid planting bananas and cucurbits near passion fruit fields.

Avocado

The avocado (*Persea americana*) is a tree native to Mexico and Central America, classified in the flowering plant family Lauraceae along with cinnamon, camphor and bay laurel. Avocado or alligator pear also refers to the fruit, botanically a large berry that contains a single seed.

Avocados are commercially valuable and are cultivated in tropical and Mediterranean climates throughout the world. They have a green-skinned, fleshy body that may be pear-shaped, egg-shaped, or spherical. Commercially, they ripen after harvesting. Trees are partially self-pollinating and often are propagated through grafting to maintain a predictable quality and quantity of the fruit.

Cultivation

Figure: *Persea americana, young avocado plant (seedling), complete with parted pit and roots*

The tree grows to 20 m (66 ft), with alternately arranged leaves 12 centimetres (4.7 in) – 25 centimetres (9.8 in) long. The flowers are inconspicuous, greenish-yellow, 5 millimetres (0.2 in) – 10 millimetres (0.4 in) wide. The pear-shaped fruit is 7 centimetres (2.8 in) – 20 centimetres (7.9 in) long, weighs between 100 grams (3.5 oz) – 1,000 grams (35 oz), and has a large central seed, 5 centimetres (2.0 in) – 6.4 centimetres (2.5 in) long.

The subtropical species needs a climate without frost and with little wind. High winds reduce the humidity, dehydrate the flowers, and affect pollination. When even a mild frost occurs, premature fruit drop may occur, although the Hass cultivar can tolerate temperatures down to –1 °C. The trees also need well-aerated soils, ideally more than 1 m deep. Yield is reduced when the irrigation water is highly saline. These soil and climate conditions are available only in a few areas of the world, particularly in southern Spain, Portugal, Morocco, Crete, the Levant, South Africa, Colombia, Peru, parts of central and northern Chile, Vietnam, Indonesia, parts of southern India, Sri Lanka, Australia, New Zealand, the Philippines, Malaysia, Central America, the Caribbean, Mexico, California, Arizona, Puerto Rico, Texas, Florida, Hawaii, Ecuador and Rwanda. Each region has different types of cultivars.

Harvest and Postharvest

Commercial orchards produce an average of seven tonnes per hectare each year, with some orchards achieving 20 tonnes per hectare. Biennial bearing can be a problem, with heavy crops in one year being followed by poor yields the next. The avocado tree does not tolerate freezing temperatures, and can be grown only in subtropical or tropical climates. There are several cold-hardy varieties planted in the region of Gainesville, Florida, which survive temperatures as low as –6.5 °C (20 °F) with only minor leaf damage.

The avocado is a climacteric fruit (the banana is another), which means it matures on the tree, but ripens off the tree. Avocados used in commerce are picked hard and green and kept in coolers at 3.3 to 5.6 °C (37.9 to 42.1 °F) until they reach their final destination. Avocados must be mature to ripen properly. Avocados that fall off the tree ripen on the ground. Generally, the fruit is picked once it reaches maturity; Mexican growers pick Hass-variety avocados when they have more than 23% dry matter, and other producing countries have similar standards. Once picked, avocados ripen in one to two weeks (depending on the cultivar) at room temperature (faster if stored with other fruits such as apples or bananas, because of the influence of ethylene gas). Some supermarkets sell pre-ripened avocados which have been treated with synthetic ethylene to hasten ripening. The use of an ethylene gas "ripening room", which is now an "industry standard," was pioneered in the 1980s by farmer Gil Henry of Escondido, California, in response to footage from a hidden supermarket camera which showed shoppers repeatedly squeezing hard, unripe avocados, putting them "back in the bin," and moving on without making a purchase. In some cases avocados can be left on the tree for several months, which is an advantage to commercial growers who seek the greatest return for their crop; but if the fruit remains unpicked for too long it falls to the ground.

Breeding

The species is only partially able to self-pollinate because of dichogamy in its flowering. This limitation, added to the long juvenile period, makes the species difficult to breed. Most cultivars are propagated via grafting, having originated from random seedling plants or minor mutations derived from cultivars. Modern breeding programmes tend to use isolation plots where the chances of cross-pollination are reduced. That is the case for programmes at the University of California, Riverside, as well as the Volcani Centre and the Instituto de Investigaciones Agropecuarias in Chile.

The avocado is unusual in that the timing of the male and female flower phases differs among cultivars. There are two flowering types, "A" and "B". "A" cultivar flowers open as female on the morning of the first day and close in late morning or early afternoon. Then they open as male in the afternoon of the second day. "B" varieties open as female on the afternoon of the first day, close in late afternoon and reopen as male the following morning.

- "A" cultivars: Hass, Gwen, Lamb Hass, Pinkerton, Reed.
- "B" cultivars: Fuerte, Sharwil, Zutano, Bacon, Ettinger, Sir Prize, Walter Hole.

Certain cultivars, such as the Hass, have a tendency to bear well only in alternate years. After a season with a low yield, due to factors such as cold (which the avocado does not tolerate well), the trees tend to produce abundantly the next season. In addition, due to environmental circumstances during some years, seedless avocados may appear on the trees. Known in the avocado industry as "cukes", they are usually discarded commercially due to their small size.

Propagation and Rootstocks

Avocados can be propagated by seed, taking roughly four to six years to bear fruit, although in some cases seedlings can take 10 years to come into bearing. The offspring is unlikely to be identical to the parent cultivar in fruit quality. Prime quality varieties are therefore propagated by grafting to rootstocks that are propagated by seed (seedling rootstocks) or by layering (clonal rootstocks). After about a year of growing in a greenhouse, the young rootstocks are ready to be grafted. Terminal and lateral grafting is normally used. The scion cultivar grows for another 6–12 months before the tree is ready to be sold. Clonal rootstocks are selected for tolerance of specific soil and disease conditions, such as poor soil aeration or resistance to the soil-borne disease (root rot) caused by *Phytophthora*.

Growing Indoors

Usually, avocados are grown from pits indoors. This is often done by removing the pit from a ripe, unrefrigerated avocado. The pit is then stabbed with three or four toothpicks, about one third of the way up. The pit is placed in a jar or vase containing tepid water. It should split in four to six weeks and yield roots and a sprout. If there is no change by this time, the avocado pit is discarded. Once the stem has grown a few inches, it is placed in a pot with soil. It should be watered every few days. Avocados have been known to grow large, so owners must be ready to repot the plant several times.

Diseases

Avocado trees are vulnerable to bacterial, viral, fungal and nutritional diseases (excesses and deficiencies of key minerals). Disease can affect all parts of the plant, causing spotting, rotting, cankers, pitting and discolouration.

A Cultivars

Choquette: A seedling from Miami, Florida, on the property of Remi Choquette. Now a favoured commercial cultivar in south Florida.

Hass: While dozens of cultivars are grown, the Hass avocado is today the most common. It produces fruit year-round and accounts for 80% of cultivated avocados in the world. All Hass avocado trees are descended from a single "mother tree" raised by a mail carrier named Rudolph Hass, of La Habra Heights, California. Hass patented the productive tree in 1935.

The "mother tree", of uncertain subspecies, died of root rot and was cut down in September, 2002. Hass trees have medium-sized (150–250 g or 5.3–8.8 oz), ovate fruit with a black, pebbled skin. The flesh has a nutty, rich flavor with 19% oil. A hybrid Guatemalan type can withstand temperatures to –1 °C (30 °F).

Gwen: A seedling bred from Hass x Thille in 1982, Gwen is higher yielding and more dwarfing than Hass in California. The fruit has an oval shape, slightly smaller than Hass (100–200 g or 3.5–7.1 oz), with a rich, nutty flavor. The skin texture is more finely pebbled than Hass, and is dull green when ripe. It is frost-hardy down to –1 °C (30 °F).

Lula: A seedling reportedly grown from a "Taft" avocado planted in Miami, Florida, on the property of George Cellon, named after Cellon's wife, Lula. It was likely a cross between Mexican and Guatemalan types. Lula was recognized for its flavor and high oil content and propagated commercially in Florida. It is also very commonly used as a rootstock for nursery production. Hardy to –4 °C (25 °F)

Pinkerton: First grown on the Pinkerton Ranch in Saticoy, California, in the early 1970s, Pinkerton is a seedling of Hass's Rincon. The large fruit has a small seed, and its green skin deepens in colour as it ripens. The thick flesh has a smooth, creamy texture, pale green colour, good flavor and high oil content. It shows some cold tolerance, to –1 °C (30 °F) and bears consistently heavy crops. A hybrid Guatemalan type, it has excellent peeling characteristics.

Reed: Developed from a chance seedling found in 1948 by James S. Reed in California, Reed has large, round, green fruit with a smooth texture and dark, thick, glossy skin. Smooth and delicate, the flesh has a slightly nutty flavor. The skin ripens green. A Guatemalan type, it is hardy to –1 °C (30 °F). Tree size is about 5 by 4 metres (16.4 by 13.1 ft).

B Cultivars

Bacon: Developed by a farmer, James Bacon, in 1954, Bacon has medium-sized fruit with smooth, green skin with yellow-green, light tasting flesh. When ripe, the skin remains green, but darkens slightly, and fruit yields to gentle pressure. It is cold-hardy down to –5 °C (23 °F).

Brogden: Possibly a cross between Mexican and West Indian types, Brogden originated as a seedling grown in Winter Haven, Florida on the property of Tom W. Brogden. The variety was recognized for its cold-hardiness to –5 °C (23 °F) and became commercially propagated as nursery-stock for home growing. It is noted for its dark purple skin at maturity.

Ettinger: A Mexican/Guatemalan cross seedling of Fuerte, this cultivar originated in Israel, and was put into production there in 1947. Mature trees tolerate four hours at –6 °C (21 °F). The fruit has a smooth, thin, green skin that does not peel easily. The flesh is very pale green.

Fuerte: A Mexican/Guatemalan cross originating in Puebla, the Fuerte earned its name, which means strong in Spanish, after it withstood a severe frost in California in 1913. Hardy to –3 °C (27 °F), it has medium-sized, pear-shaped fruit with a green, leathery, easy to peel skin. The creamy flesh of mild and rich flavour has 18% oil. The skin ripens green. Tree size is 6 by 4 metres (19.7 by 13.1 ft).

Monroe: A Guatemalan/West Indian cross that originated from a seedling grown in Homestead, Florida on the property of J.J.L. Phillips, it was patented in 1937 and became a major commercial cultivar due to its cold hardiness and production qualities. The fruit is large, averaging over two pounds (0.91 kg) in weight, has an elliptical shape, and green, glossy skin. Hardy to –3 °C (27 °F).

Sharwil: Predominantly Guatemalan, with some Mexican race genes, Sharwil was selected in 1951 by Sir Frank Sharpe at Redland Bay, southern Queensland, Australia. The name "Sharwil" is an amalgamation of Sharp and Wilson (J.C. Wilson being the first propagator). Scions were sent from Australia to Hawaii in 1966. A medium-sized fruit with rough green skin, it closely resembles the Fuerte, but is slightly more oval in shape. The fruit has greenish-yellow flesh with a rich, nutty flavor and high oil content (20–24%), and a small seed. The skin is green when ripe. It represents more than 57% of the commercial farming in Hawaii, and represents up to 20% of all avocados grown in New South Wales, Australia. It is a regular and moderate bearer with excellent quality fruit, but is sensitive to frost. Disease and pest resistance are superior to Fuerte.

Zutano: Originated by R.L. Ruitt in Fallbrook in 1926, this Mexican variety is hardy to –4 °C (25 °F). The large, pear-shaped fruit has a shiny, thin, yellow-green skin that peels moderately easily. The flesh is pale green with fibres and has a light flavor.

Other Cultivars

Other avocado cultivars include Spinks. The fruit of the cultivar Florida, grown mostly outside California, is larger and rounder, with a smooth, medium-green skin, and a less-fatty, firmer and fibrous flesh. These are occasionally marketed as low-calorie avocados. Historically attested varieties (which may or may not survive among Horticulturists) include the Challenge, Dickinson, Kist, Queen, Rey, Royal, Sharpless, and Taft.

Culinary Uses

The fruit of horticultural cultivars has a markedly higher fat content than most other fruit, mostly monounsaturated fat, and as such serves as an important staple in the diet of various groups where access to other fatty foods (high-fat meats and fish, dairy products, etc.) is limited.

A ripe avocado yields to gentle pressure when held in the palm of the hand and squeezed. The flesh is prone to enzymatic browning; it turns brown quickly after exposure to air. To prevent this, lime or lemon juice can be added to avocados after they are peeled.

The fruit is not sweet, but rich, and distinctly yet subtly flavored, and of smooth, almost creamy texture. It is used in both savory and sweet dishes, though in many countries not for both. The avocado is very popular in vegetarian cuisine, as substitute for meats in sandwiches and salads because of its high fat content.

Generally, avocado is served raw, though some cultivars, including the common Hass, can be cooked for a short time without becoming bitter. Caution should be used when cooking with untested cultivars; the flesh of some avocados may be rendered inedible by heat. Prolonged cooking induces this chemical reaction in all cultivars.

It is used as the base for the Mexican dip known as guacamole, as well as a spread on corn tortillas or toast, served with spices.

In the Philippines, Brazil, Indonesia, Vietnam, and southern India (especially the coastal Kerala and Karnataka region), avocados are frequently used for milkshakes and occasionally added to ice cream and other desserts. In Brazil, Vietnam, the Philippines and Indonesia, a dessert drink is made with sugar, milk or water, and pureed avocado.

Chocolate syrup is sometimes added. In Morocco, there is a similar chilled avocado and milk drink, that is sweetened with confectioner's sugar and hinted with orange flower water.

In Ethiopia, avocados are made into juice by mixing them with sugar and milk or water, usually served with Vimto and a slice of lemon. It is also very common to serve layered multiple fruit juices in a glass (locally called *Spris*) made of avocados, mangoes, bananas, guavas and papayas. Avocados are also used to make salads.

Avocados in savory dishes, often seen as exotic, are a relative novelty in Portuguese-speaking countries, such as Brazil, where the traditional preparation is mashed with sugar and lime, and eaten as a dessert or snack. This contrasts with Spanish-speaking countries such as Chile, Mexico or Argentina, where the opposite is true and sweet preparations are often unheard of.

In Australia and New Zealand, it is commonly served in sandwiches, sushi, on toast, or with chicken. In Ghana, it is often eaten alone in sliced bread as a sandwich. In Sri Lanka, well ripened flesh, thoroughly mashed with sugar and milk, or treacle (a syrup made from the nectar of a particular palm flower) was once a popular dessert. In Haiti it is often consumed with cassava or regular bread for breakfast.

In Mexico and Central America, avocados are served mixed with white rice, in soups, salads, or on the side of chicken and meat. In Peru, they are consumed with *tequeños* as mayonnaise, served as a side dish with *parrillas*, used in salads and sandwiches, or as a whole dish when filled with tuna, shrimp, or chicken.

In Chile, it is used as a puree-like sauce with chicken, hamburgers, and hot dogs; and in slices for celery or lettuce salads. The Chilean version of Caesar salad contains large slices of mature avocado. In Kenya and Nigeria, the avocado is often eaten as a fruit, and is eaten alone, or mixed with other fruits in a fruit salad, or as part of a vegetable salad. In Iran, it is used as a rejuvenating facial cream.

Avocado slices are frequently added to hamburgers, *tortas*, hot dogs, and *carne asada*. Avocado can be combined with eggs (in scrambled eggs, tortillas or omelettes), and is a key ingredient in California rolls and other *makizushi* ("maki", or rolled sushi).

In southern Africa, Avocado Ritz is a common dish.

In the United Kingdom, the avocado became widely available in the 1960s when it was introduced by Sainsbury's under the name 'avocado pear'.

Nutritional Value

Avocados have diverse fats. For a typical avocado:

- About 75% of an avocado's energy comes from fat, most of which (67% of total fat) is monounsaturated fat as oleic acid.
- Other predominant fats include palmitic acid and linoleic acid.
- The saturated fat content amounts to 14% of the total fat.
- Typical total fat composition is roughly (rounded to digits): 1% ù-3, 14% ù-6, 71% ù-9 (65% oleic and 6% palmitoleic), and 14% saturated fat (palmitic acid).
- On a weight basis, avocados have 35% more potassium (485 mg/100 g) than bananas (358 mg/100 g). They are rich in folic acid and vitamin K, and are good dietary sources of vitamin B6, vitamin C, vitamin E and pantothenic acid.
- Avocados have a high fibre content of 75% insoluble and 25% soluble fibre.

High avocado intake was shown in one preliminary study to lower blood cholesterol levels. Specifically, after a seven-day diet rich in avocados, mild hypercholesterolemia patients showed a 17% decrease in total serum cholesterol levels. These subjects also showed a 22% decrease in both LDL (harmful cholesterol) and triglyceride levels and 11% increase in HDL (helpful cholesterol) levels. A 2013 epidemiological report funded by the Hass Avocado Board showed that American avocado consumers had better overall diet quality, nutrient levels, and reduced risk of metabolic syndrome.

Extracts of avocado have been studied in laboratory research to assess potential for lowering risk of diabetes mellitus.

A Japanese team synthesised the four chiral components of avocado, and identified (2R, 4R)-16-heptadecene-1, 2, 4-triol as a potential antibacterial component. Due to a combination of specific aliphatic acetogenins, avocado is under preliminary research for potential anti-cancer activity.

As a houseplant

While not particularly popular, the avocado tree can be grown domestically and used as a (decorative) houseplant. The pit germinates in normal soil conditions or partially submerged in a small glass (or container) of water. In the latter method, the pit sprouts in four to six weeks, at which time it is planted in standard houseplant potting soil. The plant normally grows large enough to be prunable; however, it does not bear fruit unless it has ample sunlight. Home gardeners can graft a branch from a fruit-bearing plant to speed maturity, which typically takes four to six years to bear fruit. To obtain fresh avocado produce, however, more than one tree must be cultivated for cross-pollination.

Avocado Pest, Disease

Avocado Scab

- Sphaceloma perseae.
- Fungal disease causing fruit drop and marking of mature fruit
- Discrete spots appear on the leaves along the midrib, which may merge into star-like patterns
- As disease progresses leaves become distorted and stunted
- Symptoms on fruit appear as corky, raised brown to purplish brown spots which enlarge to form large rough areas
- Long distance dispersal likely through infested propagation material

Avocado Seed Moth

- Pest of avocados in Latin America causing large yield losses
- Early instar larvae are white but become violet in later instars
- Adult moths have a 'C' shape of black dots across the forewings
- Moths form galleries in branches causing them to wither
- Larvae penetrate fruit and destroy seed and pulp through feeding
- Frass can be seen in fruit when opened, and oozing of a white exudate may occur externally

Avocado Sunblotch

- Found sporadically in some areas of Australia
- Fruit are small and deformed with sunken pale blotches
- Discolouration and streaking occurs on stems
- A range of symptoms are seen on leaves though some infected trees remain symptomless
- Spread through infected seed or using infected scions for grafting

Avocado Thrips

- Emerging insect pest causing large economic losses in California
- Adults are straw coloured with dark lines across the abdomen
- Causes bronzing along leaf veins or over entire leaves under severe infestation

- Fruit are scarred with a characteristic 'alligator skin' appearance
- Scars elongate as fruit develops causing lower quality fruit
- Disseminated over long distances on infested avocado plant material

Laurel Wilt

- Emerging fungal disease of avocado trees in Florida
- Spread by the tiny (2 mm long) Redbay ambrosia beetle (exotic)
- Leaves wilt and foliage drop may follow
- Xylem tissue becomes dark brown or black
- Dying trees attract beetles in large numbers and small strings of wood fibre may be seen protruding from bore holes

Oriental Fruit Fly Complex

- Includes Oriental, Papaya and Carambola fruit flies
- Found in Asia, Papua New Guinea, the Pacific, South America
- Adults 6-8 mm long with a narrow brown band along edge of wings
- Abdomen has a black T-shaped mark, which is similar to a number of other endemic species
- Larval feeding can result in rotting of fruit and may cause fruit to drop

Persea Mite

- Serious pest in California with a wide host range
- Adult females are yellowish green with several dark spots in the abdominal region
- Males are smaller with yellow, pear-shaped bodies
- Nests form along midribs and veins on the underside of leaves
- Symptoms can also be seen as necrotic spots covered with a dense protective webbing

Defoliation occurs in heavy infestations with a litter of yellow-spotted green leaves under trees

Atemoya

The atemoya, *Annona* × *atemoya*, is a hybrid of two fruits – the sugar-apple (*Annona squamosa*) and the cherimoya (*A. cherimola*) – which are both native to the American tropics. This fruit is popular in Taiwan, where it is known as the "pineapple sugar apple" (óœ¨hË'æ), so is sometimes wrongly believed to be a cross between the sugar-apple and the pineapple. In Cuba, this fruit is called *anón*, and in Venezuela *chirimorinon*. In Palestine and Lebanon, the fruit is called *achta*. This is different from the Achta that is used in many Lebanese desserts, including ice cream, which is actually the skimming of fresh milk or cream.

An atemoya is normally heart-shaped or rounded, with pale-green, easily bruised, bumpy skin. Near the stem, the skin is bumpy as it is in the sugar-apple, but become smoother like the cherimoya on the bottom. The flesh is not segmented like that of the sugar-apple, bearing more similarity to that of the cherimoya. It is very juicy and smooth, tasting slightly sweet and a little tart, reminiscent of a *piña colada*. The taste also resembles vanilla from its sugar-apple parent. Many inedible, toxic, black seeds are found throughout the flesh of the atemoya. When ripe, the fruit can be scooped out of the shell and eaten chilled.

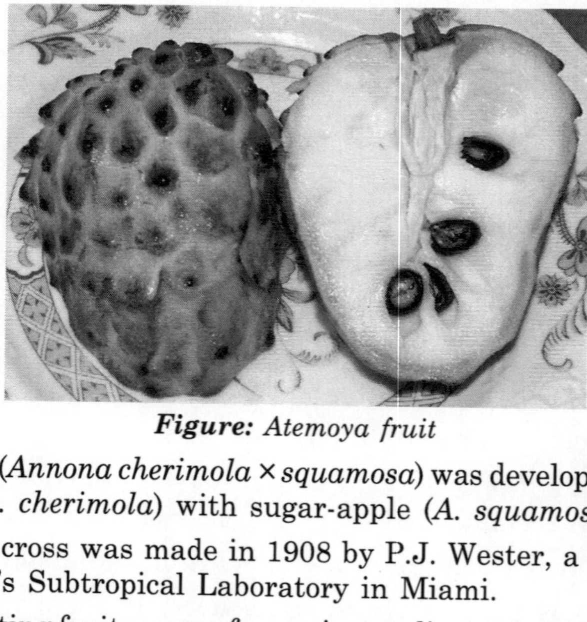

Figure: *Atemoya fruit*

Atemoya (*Annona cherimola × squamosa*) was developed by crossing cherimoya (*A. cherimola*) with sugar-apple (*A. squamosa*).

The first cross was made in 1908 by P.J. Wester, a horticulturist at the USDA's Subtropical Laboratory in Miami.

The resulting fruits were of superior quality to the sugar-apple and were given the name "atemoya", a combination of *ate*, an old Mexican name for sugar-apple, and "moya" from cherimoya.

Subsequently, in 1917, Edward Simmons at Miami's Plant Introduction Station successfully grew hybrids that survived a drop in temperature to 26.5°F, showing atemoya's hardiness derived from one of its parents, the cherimoya.

The atemoya, like other *Annona* trees, bears protogynous, hermaphroditic flowers, and self-pollination is rare. Therefore, artificial, hand pollination almost always guarantees superior quality fruits. One variety, 'Geffner', produces well without hand pollination. 'Bradley' also produces fair crops without hand pollination, but the fruit has a habit of splitting on the tree.

Atemoyas are sometimes misshapen, underdeveloped on one side, as the result of inadequate pollination.

An atemoya flower, in its female stage, opens between 2:00 and 4:00 pm; between 3:00 pm and 5:00 pm on the following afternoon, the flower converts to its male stage.

Landscaping use:

- The tree naturally grows as an espalier, which makes it a great addition to narrow planting spaces.
- In my experience it is a medium to fast-growing tree, and is said to be able to reach 30+ feet in height. However, it is easily trained with pruning in the dormant season. The wood is brittle so pruning also helps the tree to support the heavy fruit.
- The branches droop somewhat and the lowest branches often touch the ground.
- Young plants need staking.
- The leaves are green with a prominent light green central vein.
- The atemoya is semi deciduous; losing some of its leaves in late winter/early spring.

Atemoya Pollination:

- The flowers are green elongated and easily missed on casual inspection.
- An atemoya flower demonstrates its male and female parts at different times of the day.
    - o As the flower opens, the female stage is revealed around 2-4pm.
    - o The following afternoon around 3-5pm, the flower opens more to uncover the male part of the same flower.
- Atemoya's are sometimes misshapen and underdeveloped on one side due to inadequate pollination. Apparently self-pollination is rare and therefore, dedicated gardeners may self pollinate the flowers to increase fruiting and fruit size.

  However, myself, I just planted 2 atemoya next to each other and I have been getting some nice fruit without hand pollination.

Soil:

- The atemoya tree is said to be able to tolerate a wide range of soil types but prefers well draining fertile soil with neutral to slightly acidic pH.
- Water-logging is fatal.
- The atemoyas shallow root system is prone to drying out and weed competition. However, heavy mulch around the tree's root zone will help to prevent both of these issues.

Water:

- I have been watering the atemoya tree with my usual drip system irrigation 2-3 x a week in the growing season (April to January).
- The plant is prone to root rot with prolonged moist conditions.
- The trees don't seem to like the dry air of the desert where the leaves are easily damaged by dry Santa Ana winds.

Atemoya fertilization:

- In many commercial plantings, no fertilizer is applied until after the trees are well established because the young roots are very sensitive.
- After the atemoya tree is established, some have recommended the use a 6-10-16 fertilizer formula, with half of the total years dose given in the spring.
- Personally, I have been using 15-15-15 because that is what I have on hand most of the time. I spread the fertilizer around the root zone about 4 different times per year starting in the late winter and ending in mid-summer.

Temp:

- The parents of the atemoya are subtropical plants native to northwest parts of South America and traditionally grown in cool mountain altitudes at the equator.
- The plant can withstand a light frost but not much more than that. However, the plant has been reported to survive a drop in temperature to 26.5°F (-3.10°C).

Atemoya pests:

- The main pests in California are mealybugs and scale.
  - o Tanglefoot on masking tape around the trunk helps keep the ants away that bring in the mealybugs and scale.

- I have also read that you shouldn't plant this tree in old vegetable gardens, near tomatoes, eggplant or asters because atemoya are susceptible to the same soil diseases.
- Atemoyas are also susceptible to Armillaria (Oak Root Fungus) and Verticillium.
- Squirrels have also discovered the atemoya fruit. Therefore, I keep a close eye on the fruit and pick them a little early because I don't like to share with my rodent neighbours. The Havahart squirrel trap is one good-karma control option.
- I found some additional diseases to be aware of on the hort purdue website. The following are direct quotes from the site.
  - o The chalcid fly that lays eggs in the seeds and makes exit holes in the fruit permitting entrance of fungi, occasionally causes mummification of the atemoya. White wax, pink wax, and brown olive scales may be found on the foliage but are shed along with the leaves.
  - o A condition called "littleleaf" is not a disease but zinc deficiency which can be corrected by foliar spraying.
  - o Atemoyas are prone to collar rot (*Phytophthora* sp.), the first sign being an exudation of gum near the base of the trunk and on the crown roots.

Canistel

The canistel (*Pouteria campechiana*) is an evergreen tree native to southern Mexico and Central America. It is cultivated in other countries, such as Brazil, Taiwan, and Vietnam for its fruit.

The canistel grows up to 10 metres (33 ft) high, and produces orange-yellow fruit, also called yellow sapote, up to 7 centimetres (2.8 in) long, which are edible raw. Canistel flesh is sweet, with a texture often compared to that of a hard-boiled egg yolk, hence its colloquial name "eggfruit". It is closely related to the Mamey sapote and abiu.

Fruit Description

The shape and size of the fruit is highly variable, depending on the cultivar. The better selections consistently produce large ovate fruit with glossy skin weighing upwards of 14 ounces. The flesh is somewhat pasty, although the best varieties have a creamy mousse-like texture. The flavor is rich and is reminiscent of an egg-custard. The fruit may contain between one and six large brown seeds.

The canistel displays climacteric fruit ripening. A fully mature fruit shows an intense yellow skin colour. It will eventually soften and drop from the tree. Insects and birds avoid the fruit flesh, perhaps because of its astringent properties, that are much reduced in senescent fruits, but still perceptible to the human palate. Apparently mature fruits severed from the tree while still hard often fail to develop the desired climacteric changes in terms of reduced astringency and a texture reminiscent of egg yolk. This, and the fact that climacteric fruits quickly start to decay at ambient temperatures, may have contributed to the low economic importance of the canistel.

Uses

As the related lucuma, the canistel can be eaten out of hand. The ripe fruit has been made into jam, marmalade, pancakes, and flour. The ripe flesh is blended with milk and other ingredients to make a shake, and pureed it is sometimes added to custards or used in making ice cream.

The wood of the tree is occasionally used in construction where it is available, especially as planks or rafters. In its native range, it has been a source of latex used to adulterate chicle.

Etymology

Its binomial name is derived from the Mexican town of Campeche, where it is native.

In the numerous countries where it is cultivated or sold, it is known by many vernacular names; canistel is common, as are variations on egg-fruit and names referring to its yellow colour. In the Philippines it is called *chesa* or *tiessa* or *tiesa*. In Sri Lanka this fruit is known as *Laulu*, *Lavulu* or *Lawalu*. In Thailand it is known by different traditional popular names such as *Lamut Khamen* or *Tho Khamen* , attributing a hypothetical Cambodian origin to this fruit. Currently those names are officially discouraged and the name *Mon Khai*, *Tiesa*, *Khai* meaning "egg", is favoured.

The plant's name in the Vietnamese language is *cây tréng gà* ("chicken egg" plant) because of the fruit's appearance. It also has the Vietnamese name *lekima*. This is unusual because Vietnamese is a tonal, isolating language whose morphemes all consist of a single syllable. It appears that this name derives from the word *lucuma*.

Though relatively rare in East Africa, they can be found and in the Swahili language, the fruit is confusingly named "Zaituni" which is the same word used to refer to Olives.

Carambola

Carambola, also known as starfruit, is the fruit of *Averrhoa carambola*, a species of tree native to the Philippines, Indonesia, Malaysia, India, Bangladesh and Sri Lanka. The fruit is popular throughout Southeast Asia, the South Pacific and parts of East Asia. The tree is also cultivated throughout non-indigenous tropical areas, such as in Latin America, the Caribbean, and the southern United States.

The fruit has distinctive ridges running down its sides (usually five, but can sometimes vary); in cross-section, it resembles a star, hence its name. The entire fruit is edible and is usually eaten out of hand. They may also be used in cooking, and can be made into relishes, preserves, and juice drinks.

The original range of *Averrhoa carambola* is unknown today. It is believed that it may have originated from Sri Lanka or Moluccas, Indonesia; but has been cultivated in the Indian Subcontinent and Southeast Asia for hundreds of years. They remain a local favourite in those areas but have also recently gained popularity in parts of East Asia and Queensland, Australia; as well as in the Pacific Islands, particularly Tahiti, New Caledonia, Papua New Guinea, Hawaii, and Guam. They are cultivated commercially in India, Southeast Asia, southern China, Taiwan, and Florida. They are also grown in Nicaragua, Costa Rica, Panama, Colombia, Ecuador, Peru, Brazil, Jamaica, Haiti, the Dominican Republic, Puerto Rico, Trinidad, Mexico, Guyana and parts of Africa. In other areas they are usually grown as ornamentals, rather than for consumption.

Description

The fruit is about 2 to 6 inches (5.1 to 15.2 cm) in length and is an oval shape. It usually has five prominent longitudinal ridges, but in rare instances it can have as little as 4, or as many as 8 prominent longitudinal ridges. In cross section, it resembles a star. The skin is thin, smooth, and waxy and turns a light to dark yellow when ripe. The flesh is translucent and light yellow to yellow in colour. Each fruit can have 10 to 12 flat light brown seeds about 0.25 to 0.5 in (0.64 to 1.27 cm) in width and enclosed in gelatinous aril. Once removed from the fruit, they lose viability within a few days.

Like the closely related bilimbi, there are two main types of carambola, the small sour (or tart) type and the larger sweet type. The sour varieties have a higher oxalic acid content than the sweet type. A number of cultivars have been developed in recent years. The most common cultivars grown commercially include the sweet types 'Arkin' (Florida), 'Dah Pon' (Taiwan), 'Fwang Tung' (Thailand), 'Maha' (Malaysia), and 'Demak' (Indonesia); and the sour types 'Golden Star', 'Newcomb', 'Star King', and 'Thayer' (all from Florida). Some of the sour varieties like 'Golden Star' can become sweet if allowed to ripen.

Gastronomy

The entire fruit is edible, including the slightly waxy skin. The flesh is crunchy, firm, and extremely juicy. It does not contain fibres and has a texture similar in consistency to that of grapes. Carambolas are best consumed shortly after they ripen, when they are yellow with a light shade of green or just after all traces of green have disappeared. They will also have brown ridges at the edges and feel firm. Fruits picked while still slightly green will turn yellow in storage at room temperature, but will not increase in sugar content. Overripe carambola will be yellow with brown spots and can become blander in taste and soggier in consistency.

Ripe sweet type carambolas are sweet without being overwhelming as they rarely have more than 4% sugar content. They have a tart, sour undertone, and an oxalic acid odor. The taste is difficult to compare, but it has been likened to a mix of apple, pear, grape, and citrus family fruits. Unripe starfruits are firmer and sour, and taste like green apples. Ripe carambolas may also be used in cooking. In Southeast Asia, they are usually stewed in cloves and sugar, sometimes with apples. In China, they are cooked with fish. In Australia, they may be cooked as a vegetable, pickled, or made into jams. In Jamaica they are sometimes dried.

Unripe and sour type carambolas can be mixed with other chopped spices to make relishes in Australia. In the Philippines, unripe carambolas are eaten dipped in rock salt. In Thailand, they are cooked together with shrimp.

The juice from carambolas is also used in iced drinks, particularly the juice of the sour varieties. In Hawaii they are used to make sherbet, while in the Philippines they can be used as seasoning. In India, the juice is bottled for drinking.

Health

Benefits: Carambola is rich in antioxidants, potassium, and vitamin C; and low in sugar, sodium, and acid. It is also a potent source of both primary and secondary polyphenolic antioxidants. *Averrhoa carambola* has both antioxidant and antimicrobial activities. Scavenging of nitric oxide (NO) by the fruit extract is dependent on concentration and stage of ripening. Extracts showed antimicrobial activity against *E. coli*, *Klebsiella* spp., *Staphylococcus aureus*, and *Pseudomonas aeruginosa*.

Figure: *Carambolas in varying stages of ripeness*

Risks: Carambolas contains oxalic acid, which has been considered harmful to individuals suffering from kidney failure, kidney stones, or those under kidney dialysis treatment. Consumption by those with kidney failure can produce hiccups, vomiting, nausea, and mental confusion. Fatal outcomes have been documented in some patients. Recent research has however, identified another compound, caramboxin, which is structurally similar to phenylalanine, to be responsible for the observed effects.

Drug Interactions

Like the grapefruit, carambola is considered to be a potent inhibitor of seven cytochrome P450 isoforms. These enzymes are significant in the first-pass elimination of many medicines, and, thus, the consumption of carambola or its juice in combination with certain medications can significantly increase their effective dosage within the body. Research into grapefruit juice has identified a number of common medications affected, including statins, which are commonly used to treat cardiovascular illness, and benzodiazepines (a tranquilizer family including diazepam).

Cultivation

The carambola is a tropical and subtropical fruit. It can be grown at up to 4,000 feet (1,200 m) in elevation. It prefers full sun exposure, but requires enough humidity and a total of 70 inches or more of rainfall a year. It does not have a soil type preference, but it requires good drainage. Carambola trees are planted at least 20 feet (6.1 m) from each other and typically are fertilized three times a year. The tree grows rapidly and typically produces fruit at four or five years of age. The large amount of rain during spring actually reduces the amount of fruit, but, in ideal conditions, carambola can produce from 200 to 400 pounds (91 to 181 kg) of fruit a year. The carambola tree flowers throughout the year, with main fruiting seasons from April to June and October to December in Malaysia, for example, but fruiting also occurs at other times in some other locales, such as South Florida.

Major pests are fruit flies, fruit moths, ants, and birds. Crops are also susceptible to frost, especially in the United States and in the Philippines. Top producers of carambola in the world market include Australia, Guyana, India, Israel, Malaysia, the Philippines, Taiwan, and the United States. Malaysia is a global leader in starfruit production by volume and ships the product widely to Asia and Europe. Due to concerns over pests and pathogens, however, whole starfruits cannot yet be imported to the US from Malaysia under current Food and Drug Administration regulations. In the United States, carambolas are grown in tropical and semitropical areas, including Texas, Florida and Hawaii.

In the United States, commercial cultivation and broad consumer acceptance of the fruit only dates to the 1970s. That acceptance is attributable to Morris Arkin, a backyard horticulturalist, from Coral Gables, Florida. During the late 1960s, Arkin began cultivating plants and trees in his backyard, eventually developing a kind of carambola, or star fruit, that became commercially viable and was named after him. Until the early 1970s, carambola had been grown only as specimen trees in botanical gardens and experiment stations and as a curiosity in home landscapes. However, because of its attractive star shape when cut in cross-section and yellow to golden colour, it began to grow in popularity. Fruit from early introductions were however, sour and sometimes considered unpalatable. This limited market and public acceptance, inhibiting development and expansion of carambola as a commercial fresh fruit. Arkin cultivated the 'Arkin' variety – a sweet carambola with good handling characteristics – in the mid to late 1970s. Soon afterwards, the limited commercial area of carambola under cultivation in south Florida (4 to 12 ha) was top-worked to 'Arkin' and this new cultivar led to a rapid increase in consumer demand for the fruit which further stimulated interest in establishing new commercial plantings. Today, the 'Arkin' variety represents 98% of the current acreage in South Florida.

Other Uses

The trees are also grown as ornamentals for their abundant brightly coloured and unusually shaped fruits, as well as for their attractive dark green leaves and their lavender to pink flowers.

Like the bilimbi, the juice of the more acidic sour types can be used to clean rusty or tarnished metal (especially brass) as well as bleach rust stains from cloth. They may also be used as a mordant in dyeing.

The Coconut Palm

The coconut palm (*Cocos nucifera*), is a member of the family Arecaceae (palm family). It is the only accepted species in the genus *Cocos*. The term coconut can refer to the entire coconut palm, the seed, or the fruit, which, botanically, is a drupe, not a nut. The spelling cocoanut is an archaic form of the word. The term is derived from 16th-century Portuguese and Spanish *coco*, meaning "head" or "skull", from the three indentations on the coconut shell that resemble facial features.

Found throughout the tropic and subtropic area, the coconut is known for its great versatility as seen in the many uses of its different parts. Coconuts are part of the daily diets of many people. Coconuts are different from any other fruits because they contain a large quantity of "water" and when immature they are known as tender-nuts or jelly-nuts and may be harvested for drinking. When mature, they still contain some water and can be used as seednuts or processed to give oil from the kernel, charcoal from the hard shell and coir from the fibrous husk. The endosperm is initially in its nuclear phase suspended within the coconut water. As development continues, cellular layers of endosperm deposit along the walls of the coconut, becoming the edible coconut "flesh". When dried, the coconut flesh is called copra. The oil and milk derived from it are commonly used in cooking and frying; coconut oil is also widely used in soaps and cosmetics. The clear liquid coconut water within is drinkable. The husks and leaves can be used as material to make a variety of products for furnishing and decorating. It also has cultural and religious significance in many societies that use it.

Description

Plant: *Cocos nucifera* is a large palm, growing up to 30 m (98 ft) tall, with pinnate leaves 4–6 m (13–20 ft) long, and pinnae 60–90 cm long; old leaves break away cleanly, leaving the trunk smooth. Coconuts are generally classified into two general types: tall and dwarf. On very fertile land, a tall coconut palm tree can yield up to 75 fruits per year, but more often yields less than 30, mainly due to poor cultural practices. In recent years, improvements in cultivation practices and breeding have produced coconut trees that can yield more.

Fruit: Botanically, the coconut fruit is a drupe, not a true nut. Like other fruits, it has three layers: the exocarp, mesocarp, and endocarp. The exocarp and mesocarp make up the "husk" of the coconut. Coconuts sold in the shops of nontropical countries often have had the exocarp (outermost layer) removed. The mesocarp is composed of a fibre, called coir, which has many traditional and commercial uses. The shell has three germination pores (stoma) or "eyes" that are clearly visible on its outside surface once the husk is removed.

A full-sized coconut weighs about 1.44 kg (3.2 lb). It takes around 6,000 full-grown coconuts to produce a tonne of copra.

Roots: Unlike some other plants, the palm tree has neither a tap root nor root hairs, but has a fibrous root system.

The coconut palm root system consists of an abundance of thin roots that grow outward from the plant near the surface. Only a few of the roots penetrate deep into the soil for stability. The type of root system is known as fibrous or adventitious, and is a characteristic of grass species. Other types of large trees produce a single downward-growing tap root with a number of feeder roots growing from it.

Coconut palms continue to produce roots from the base of the stem throughout its life. The number of roots produced depends on the age of the tree and the environment, with more than 3,600 roots possible on a tree that's 60 to 70 years old.

Roots are usually less than about 3 inches in diametre and uniformly thick from the tree trunk to the root tip.

Inflorescence

The palm produces both the female and male flowers on the same inflorescence; thus, the palm is monoecious. Other sources use the term polygamomonoecious. The female flower is much larger than the male flower. Flowering occurs continuously. Coconut palms are believed to be largely cross-pollinated, although some dwarf varieties are self-pollinating.

Etymology

One of the earliest mentions of the coconut dates back to the One Thousand and One Nights story of Sinbad the Sailor; he is known to have bought and sold coconuts during his fifth voyage. *Tenga*, its Malayalam name, was used in the detailed description of coconut found in *Itinerario* by Ludovico di Varthema published in 1510 and also in the later *Hortus Indicus Malabaricus*. Even earlier, it was called *nux indica*, a name used by Marco Polo in 1280 while in Sumatra, taken from the Arabs who called it ÌæÒ åäÏí *jawz hindî*. Both names translate to "Indian nut". In the earliest description of the coconut palm known, given by Cosmos of Alexandria in his *Topographia Christiana* written about 545 AD, there is a reference to the argell tree and its drupe.

Historical evidence favours the European origin of the name "coconut", for no name is similar in any of the languages of India, where the Portuguese first found the fruit; and indeed Barbosa, Barros, and Garcia, in mentioning the Malayalam name *tenga*, and Canarese *narle*, expressly say, "we call these fruits *quoquos*", "our people have given it the name of coco", and "that which we call coco, and the Malabars temga".

The OED states: "Portuguese and Spanish authors of the 16th c. agree in identifying the word with Portuguese and Spanish *coco* "grinning face, grin, grimace", also "bugbear, scarecrow", cognate with *cocar* "to grin, make a grimace"; the name being said to refer to the face-like appearance of the base of the shell, with its three holes. According to Losada, the name came from Portuguese explorers, the sailors of Vasco da Gama in India, who first brought them to Europe. The coconut shell reminded them of a ghost or witch in Portuguese folklore called *coco* (also *côca*).

Natural Habitat

The coconut palm thrives on sandy soils and is highly tolerant of salinity. It prefers areas with abundant sunlight and regular rainfall (1500 mm to 2500 mm annually), which makes colonizing shorelines of the tropics relatively straightforward. Coconuts also need high humidity (70–80%+) for optimum growth, which is why they are rarely seen in areas with low humidity, like the southeastern Mediterranean or Andalusia, even where temperatures are high enough (regularly above 24 °C or 75.2 °F).

Coconut palms require warm conditions for successful growth, and are intolerant of cold weather. Optimum growth is with a mean annual temperature of 27 °C (81 °F), and growth is reduced below 21 °C (70 °F). Some seasonal variation is tolerated, with good growth where mean summer temperatures are between 28 and 37 °C (82 and 99 °F), and survival as long as winter temperatures are above 4–12 °C (39–54 °F); they will survive brief drops to 0 °C (32 °F). Severe frost is usually fatal, although they have been known to recover from temperatures of "4 °C (25 °F). They may grow but not fruit properly in areas with insufficient warmth, such as Bermuda.

The conditions required for coconut trees to grow without any care are:

- Mean daily temperature above 12–13 °C (53.6–55.4 °F) every day of the year
- Mean annual rainfall above 1,000 mm (39.37 in)
- No or very little overhead canopy, since even small trees require direct sun

The main limiting factor for most locations which satisfy the rainfall and temperature requirements is canopy growth, except those locations near coastlines, where the sandy soil and salt spray limit the growth of most other trees.

Diseases

Coconuts are susceptible to the phytoplasma disease lethal yellowing. One recently selected cultivar, the Maypan, has been bred for resistance to this disease.

Pests

The coconut palm is damaged by the larvae of many Lepidoptera (butterfly and moth) species which feed on it, including *Batrachedra* spp.: *B. arenosella*, *B. atriloqua* (feeds exclusively on *C. nucifera*), *B. mathesoni* (feeds exclusively on *C. nucifera*), and *B. nuciferae*.

Brontispa longissima (coconut leaf beetle) feeds on young leaves, and damages seedlings and mature coconut palms. In 2007, the Philippines imposed a quarantine in Metro Manila and 26 provinces to stop the spread of the pest and protect the $800 million Philippine coconut industry.

The fruit may also be damaged by eriophyid coconut mites (*Eriophyes guerreronis*). This mite infests coconut plantations, and is devastating: it can destroy up to 90% of coconut production. The immature seeds are infested and desapped by larvae staying in the portion covered by the perianth of the immature seed; the seeds then drop off or survive deformed. Spraying with wettable sulfur 0.4% or with neem-based pesticides can give some relief, but is cumbersome and labour-intensive.

In Kerala, the main coconut pests are the coconut mite, the rhinoceros beetle, the red palm weevil and the coconut leaf caterpillar. Research on this topic has as of 2009 produced no results, and researchers from the Kerala Agricultural University and the Central Plantation Crop Research Institute, Kasaragode are still searching for a cure. The Krishi Vigyan Kendra, Kannur under Kerala Agricultural University has developed an innovative extension approach called compact area group approach (CAGA) to combat coconut mites.

Cultivation

Coconut palms are grown in more than 80 countries of the world, with a total production of 61 million tonnes per year. Coconut trees are very hard to establish in dry climates, and cannot grow there without frequent irrigation; in drought conditions, the new leaves do not open well, and older leaves may become desiccated; fruit also tends to be shed.

The extent of cultivation in the tropics is threatening a number of habitats, such as mangroves; an example of such damage to an ecoregion is in the Petenes mangroves of the Yucatán.

Harvesting

In some parts of the world (Thailand and Malaysia), trained pig-tailed macaques are used to harvest coconuts. Training schools for pig-tailed macaques still exist both in southern Thailand, and in the Malaysian state of Kelantan. Competitions are held each year to find the fastest harvester.

Commercial, Iindustrial and Household Use

Cultivars: Coconut has a number of commercial and traditional cultivars. They can be sorted mainly into tall cultivars, dwarf cultivars and hybrid cultivars (hybrids between talls and dwarfs). Some of the dwarf cultivars such as *Malayan dwarf* has shown some promising resistance to lethal yellowing while other cultivars such as *Panama tall* is highly affected by the same plant disease. Some cultivars are more drought resistant such as *West coast tall* (India) while others such as *Hainan Tall* (China) are more cold tolerant. Other aspects such as seed size, shape and weight and copra thickness are also important factors in the selection of new cultivars. Some cultivars such as *Fiji dwarf* form a large bulb at the lower stem and others are cultivated to produce very sweet coconut water with orange coloured husks (king coconut) used entirely in fruit stalls for drinking (Sri Lanka, India).

Coir: Coir (the fibre from the husk of the coconut) is used in ropes, mats, door mats, brushes, sacks, caulking for boats, and as stuffing fibre for mattresses. It is used in horticulture in potting compost, especially in orchid mix.

Coconut Fronds: The stiff mid-ribs of coconut leaves are used for making brooms in India, Indonesia (*sapu lidi*), Malaysia, the Maldives and the Philippines (*walis tingting*). The green of the leaves (lamina) are stripped away, leaving the veins (wood-like, thin, long strips) which are tied together to form a broom or brush. A long handle made from some other wood may be inserted into the base of the bundle and used as a two-handed broom. The leaves also provide material for baskets that can draw well water and for roofing thatch; they can be woven into mats, cooking skewers, and kindling arrows, as well. Two leaves (especially the younger, yellowish shoots) woven into a tight shell the size of the palm are filled with rice and cooked to make *ketupat*. Dried coconut leaves can be burned to ash, which can be harvested for lime. In India, the woven coconut leaves are used as *pandals* (temporary sheds) for marriage functions especially in the states of Kerala, Karnataka, and Tamil Nadu.

Copra: Copra is the dried meat of the seed and after processing produces coconut oil and coconut meal. Coconut oil, aside from being used in cooking as an ingredient and for frying, is used in soaps, cosmetics, hair-oil, and massage oil. Coconut oil is also a main ingredient in Ayurvedic oils. In Vanuatu coconut palms for copra production are generally spaced 9 metres apart, allowing a tree density of 100–160 trees per hectare.

Husks and Shells

The husk and shells can be used for fuel and are a source of charcoal. Activated carbon manufactured from coconut shell is considered extremely effective for the removal of impurities. The coconut's obscure origin in foreign lands led to the notion of using cups made from the shell to neutralise poisoned drinks. The cups were frequently engraved and decorated with precious metals.

A dried half coconut shell with husk can be used to buff floors. It is known as a *bunot* in the Philippines and simply a "coconut brush" in Jamaica. The fresh husk of a brown coconut may serve as a dish sponge or body sponge. *Tempurung* as the shell is called in the Malay language can be used as a soup bowl and—if fixed with a handle—a ladle. In India, coconut shells are also used as bowls and in the manufacture of various handicrafts, including buttons carved from dried shell. Coconut buttons are often used for Hawaiian aloha shirts. In Thailand, the coconut husk is used as a potting medium to produce healthy forest tree saplings. The process of husk extraction from the coir bypasses the retting process, using a custom-built coconut husk extractor designed by ASEAN–Canada Forest Tree Seed Centre (ACFTSC) in 1986. Fresh husks contains more tannin than old husks. Tannin produces negative effects on sapling growth. In parts of South India, the shell and husk are burned for smoke to repel mosquitoes.

Half coconut shells are used in theatre Foley sound effects work, banged together to create the sound effect of a horse's hoofbeats. Dried half shells are used as the bodies of musical instruments, including the Chinese *yehu* and *banhu*, along with the Vietnamese *đàn gáo* and

Arabo-Turkic *rebab*. In the Philippines, dried half shells are also used as a music instrument in a folk dance called *maglalatik*.

In World War II, coastwatcher scout Biuki Gasa was the first of two from the Solomon Islands to reach the shipwrecked and wounded crew of Motor Torpedo Boat PT-109 commanded by future U.S. president John F. Kennedy. Gasa suggested, for lack of paper, delivering by dugout canoe a message inscribed on a husked coconut shell. This coconut was later kept on the president's desk, and is now in the John F. Kennedy Library.

Coconut Trunk

Coconut trunks are used for building small bridges and huts; they are preferred for their straightness, strength, and salt resistance. In Kerala, coconut trunks are used for house construction. Coconut timber comes from the trunk, and is increasingly being used as an ecologically sound substitute for endangered hardwoods. It has applications in furniture and specialized construction, as notably demonstrated in Manila's Coconut Palace.

Hawaiians hollowed the trunk to form drums, containers, or small canoes. The "branches" (leaf petioles) are strong and flexible enough to make a switch. The use of coconut branches in corporal punishment was revived in the Gilbertese community on Choiseul in the Solomon Islands in 2005.

Coconut Roots

The roots are used as a dye, a mouthwash, and a medicine for diarrhea and dysentery. A frayed piece of root can also be used as a toothbrush.

Use in Beauty Products

Coconuts are used in the beauty industry in moisturisers and body butters because coconut oil, due to its chemical structure, is readily absorbed by the skin. The coconut shell may also be ground down and added to products for exfoliation of dead skin. Coconut is also a source of lauric acid, which can be processed in a particular way to produce sodium lauryl sulfate, a detergent used in shower gels and shampoos. The nature of lauric acid as a fatty acid makes it particularly effective for creating detergents and surfactants.

Guava

Guavas are plants in the Myrtle family (Myrtaceae) genus *Psidium*, which contains about 100 species of tropical shrubs and small trees. They are native to Mexico, Central America, and northern South America.

Guavas are now cultivated and naturalized throughout the tropics and subtropics in Africa, South Asia, Southeast Asia, the Caribbean, subtropical regions of North America, Hawaii, New Zealand, Australia and Spain.

Types

The most frequently eaten species, and the one often simply referred to as "the guava", is the Apple Guava (*Psidium guajava*).. Guavas are typical Myrtoideae, with tough dark leaves that are opposite, simple, elliptic to ovate and 5–15 centimetres (2.0–5.9 in) long. The flowers are white, with five petals and numerous stamens.

The genera *Accara* and *Feijoa* (= *Acca*, Pineapple Guava) were formerly included in *Psidium*.

Common Names

The term "guava" appears to derive from Arawak *guayabo* "guava tree", via the Spanish *guayaba*. It has been adapted in many European and Asian languages, having a similar form.

Another term for guavas is *pera*, derived from pear. It is common around the western Indian Ocean and probably derives from Spanish or Portuguese. In the Indian subcontinent and Middle-East, guava is also called *amrood*, possibly a variant of *armoot* meaning "pear" in Arabic and Turkish languages.

In Egypt, it is called *gawafa*.

Ecology

Psidium species are used as food plants by the caterpillars of some Lepidoptera, mainly moths like the Ello Sphinx (*Erinnyis ello*), *Eupseudosoma aberrans*, *E. involutum*, and *Hypercompe icasia*. Mites like *Pronematus pruni* and *Tydeus munsteri* are known to parasitize the Apple Guava (*P. guajava*) and perhaps other species. The bacterium *Erwinia psidii* causes rot diseases of the Apple Guava.

The fruit is not only relished by humans, but by many mammals and birds as well. The spread of introduced guavas owes much to this fact, since animals eat the fruit and disperse the seeds in their droppings.

In several tropical regions, including Hawaii and Florida, some species (namely Strawberry Guava, *P. littorale*, and to a lesser extent Apple Guava) have become invasive species. On the other hand, several species have become very rare due to habitat destruction and at least one (Jamaican Guava, *P. dumetorum*), is already extinct.

Guava wood is used for meat smoking in Hawaii and is used at barbecue competitions across the United States. In Cuba and Mexico the leaves are used in barbecues.

Fruit

Guava fruit, usually 4 to 12 centimetres (1.6 to 4.7 in) long, are round or oval depending on the species. They have a pronounced and typical fragrance, similar to lemon rind but less sharp. The outer skin may be rough, often with a bitter taste, or soft and sweet. Varying between species, the skin can be any thickness, is usually green before maturity, but becomes yellow, maroon, or green when ripe. The pulp inside may be sweet or sour, and off-white ("white" guavas) to deep pink ("red" guavas). The seeds in the central pulp vary in number and hardness, depending on species.

Range

Guavas are cultivated in many tropical and subtropical countries. Several species are grown commercially; apple guava and its cultivars are those most commonly traded internationally.

Mature trees of most species are fairly cold-hardy and can survive temperatures slightly colder than 25 °F (–4 °C) for short periods of time, but younger plants will likely freeze to the ground. Guavas are grown in South Florida as far north as Sarasota, on the west coast, and Fort Pierce, on the east coast. However, they are a primary host of the Caribbean Fruit Fly and must be protected against infestation in areas of Florida where this pest is present.

Guavas are also of interest to home growers in temperate areas. They are one of the few tropical fruits that can grow to fruiting size in pots indoors. When grown from seed, guavas can bear fruit as soon as two years, or as long as eight years.

Culinary Uses

In Mexico, the guava agua fresca beverage is popular. The entire fruit is a key ingredient in punch, and the juice is often used in culinary sauces (hot or cold), as well as artisan candies, dried snacks, fruit bars, desserts, or dipped in Chamoy. *Pulque de Guava* is a popular blend of the native alcoholic beverage.

In many countries, guava is eaten raw, typically cut into quarters or eaten like an apple, whereas in other countries it's eaten with a pinch of salt and pepper, cayenne powder or a mix of spices (masala). It is known as the winter national fruit of Pakistan. In the Philippines, ripe guava is used in cooking sinigang. Guava is a popular snack in Taiwan, sold on many street corners and night markets during hot weather, accompanied by packets of dried plum powder mixed with sugar and salt for dipping. In east Asia, guava is commonly eaten with sweet and sour dried plum powder mixtures. Guava juice is popular in many countries. The fruit is also often prepared in fruit salads. Because of its high level of pectin, guavas are extensively used to make candies, preserves, jellies, jams, and marmalades (such as Brazilian *goiabada* and Colombian and Venezuelan *bocadillo*), and also for juices and aguas frescas or may be used in a marmalade jam on toast.

Red guavas can be used as the base of salted products such as sauces, substituting for tomatoes, especially to minimize acidity. A drink may be made from an infusion of guava fruits and leaves which in Brazil is called chá-de-goiabeira, i.e. "tea" of guava tree leaves, considered medicinal.

Nutritional Value

Guavas are rich in dietary fibre and vitamin C, with moderate levels of folic acid. Having a generally broad, low-calorie profile of essential nutrients, a single common guava (*P. guajava*) fruit contains about four times the amount of vitamin C as an orange. However, nutrient content varies across guava cultivars. Although the strawberry guava (*P. littorale* var. *cattleianum*) has about 25% of the amount found in more common varieties, its total vitamin C content in one serving (90 mg) still provides 100% of the Dietary Reference Intake.

Guavas contain both carotenoids and polyphenols like (+)-gallocatechin, guaijaverin, leucocyanidin and amritoside–the major classes of antioxidant pigments – giving them relatively high potential antioxidant value among plant foods. As these pigments produce the fruit skin and flesh colour, guavas that are red-orange have more pigment content as polyphenol, carotenoid and pro-vitamin A, retinoid sources than yellow-green ones.

Potential Medical Uses

Since the 1950s, guavas – particularly the leaves – have been the subject for diverse research on their constituents, pharmacological properties and history in folk medicine. Most research, however, has been conducted on apple guava (*P. guajava*), with other species remaining unstudied. From preliminary medical research in laboratory models, extracts from apple guava leaves or bark are implicated in therapeutic mechanisms against cancer, bacterial infections, inflammation and pain. Essential oils from guava leaves display anti-cancer activity in vitro.

Guava leaves are used in folk medicine as a remedy for diarrhea and, as well as the bark, for their supposed antimicrobial properties and as an astringent. Guava leaves or bark are used in traditional treatments against diabetes. In Trinidad, a tea made from young leaves is used for diarrhea, dysentery and fever.

Pests of Guava

Fruit Borers

Congethes (=Dichocrocis) punctiferalis

Symptoms of Damage

Caterpillar bores into young fruits

Feeds on internal contents (pulp and seeds)Dry up and fall off in without ripening

Identification of pest

- Larva - Pale greenish with pink tinge and fine hairs with dark head
- Adult - Yellowish moth with black spots on the wing and body.

Management:

- Collect and destroy damaged fruits
- Clean cultivation as weed plants serve as alternate hosts
- Use light trap @ 1/ ha to monitor the activity of adults
- Insecticides: malathion 50 EC 0.1% two rounds, one at flower formation and next at fruit set.

Fruit borers: *Deudorix (Virachola) isocrates*

Symptoms of damage:

- Caterpillar bores into young fruits
- Feeds on internal contents (pulp and seeds)

Identification of pest:

- Larvae - dark brown, short and stout, covered with short hairs
- Adult - blusih brown butterfly. Female – V shaped patch on forewing

Management:

- Collect and destroy damaged fruits
- Clean cultivation as weed plants serve as alternate hosts
- Use light trap @ 1/ ha to monitor the activity of adults
- Insecticides: malathion 50 EC 0.1% two rounds, one at flower formation and next at fruit set.

Fruit borer, *Rapala varuna,*

Symptoms of damage:

- Caterpillar found on flower buds and fruit

Identification of pest:

- Adult – metallic red coloured butterfly

Management:

- Collect and destroy damaged fruits
- Clean cultivation as weed plants serve as alternate hosts
- Use light trap @ 1/ ha to monitor the activity of adults
- Insecticides: malathion 50 EC 0.1% two rounds, one at flower formation and next at fruit set.

Guava fruit fly: *Bactrocera diversus*

Symptoms of damage:

- Adults and maggots attack semi – ripe fruits
- Oviposition punctures on fruits
- Maggots destroy and convert pulp into a bad smelling
- Discoloured semi liquid mass

Identification of pest:

- Adult - Brown or dark brown with hyaline wings and yellow legs

Management:

- Collect and destroy fallen and infested fruits
- Summer ploughing to expose pupa
- Use methyl eugenol lure trap (25/ha) to monitor and kill adults of fruit flies
- Prepare methyl eugenol and malathion 50 EC mixture at 1:1 ratio(take 10 ml mixture/ trap)
- Inseciticides: malathion 50 EC 0.05%
- Bait spray combining molasses or jaggery 10g/l and one of the insecticides,
    - o fenthion 100 EC 1ml/l
    - o malathion 50 EC 2 ml/l
    - o dimethoate 30 EC 1ml/lit, two rounds at fortnight interval before ripening of fruits
    - o Field release of parasitoids such as
        - – Opius compensatus
        - – *Spalangia philippinensis*,
        - – parasitoid wasp *Diachasmimorpha kraussi*

Bark eating caterpillars, *Indarbella* sp

Symptoms of damage:

- Young trees may succumb to the attack.
- Caterpillars bore into the trunk or junction of branches.
- Presence of gallery made out of silk and frass.
- Caterpillars remain hidden in the tunnel during day time come out at night and feed on the bark.

Identification of pest:

- Adult is stout yellowish –brown moth with brown wavy markings on the forewings.
- Males are smaller than the females.

Management:

- Remove and destroy dead and severely affected branches of the tree
- Remove alternate host, silk cotton and other hosts
- Swab Coal tar + Kerosene @ 1:2 or Carbaryl 50 WP 20 g / l of water on the basal portion of the trunk up to 3 feet height
- Scraping the loose bark to prevent oviposition by adult beetles.
- Padding with monocrotophos 36 WSC 10 ml in 2.5 cm /tree soaked in absorbent cotton.
- If infestations are severe then apply the copper oxychloride paste on the trunk of the tree.
- Hook out the grub from the bore hole
- After apply monocrotophos 36 WSC 10 to 20 ml/ hole
- One celphos tablet (3 g aluminum phosphide) per hole
- Apply carbofuran 3G 5 g per hole and plug with mud.

Sap Feeders

Tea mosquito bug, *Helopeltis antonii*

Symptoms of damage:

- Nymphs and adults make punctures on petiole, tender shoots and fruits
- Brownish – black necrotic patches develop on foliage
- Elongate streaks and patches develop on shoots
- Corky scab formation on fruits

Identification of pest:

- Nymph and adult: Reddish brown, elongate bug with black head, red thorax
- black and white abdomen

Management:

- Collect and destroy the damaged plant parts
- Insecticides: malathion 50 EC 0.2% at bimonthly interval at the time of flowering

Green Scale, *Coccus viridis*

Symptoms of damage:

- Nymphs and adults suck the sap from leaves
- Yellowing of leaves.

Identification of pest:

- Nymph: Pale lemon yellow.
- Adult – green colour, flat and oval shaped soft scale

Management:

- Prune and destroy the infested shoots at initial stage of infestation.
- Spray monocrotophos 1ml/lit
- After two weeks release 20 predator beetles viz.,*C. montrouzieri*beetle per tree.
- Encourage parasitoids viz., *Aneristis* sp *Coccophgagus cowperi*, C. *bogoriensis*

Tailed mealy bug, *Ferrisa virgata*

Symptoms of damage:

- Premature dropping of fruit.

Identification of pest:

- Nymph - Yellowish to pale white.
- Adult - Females apterous, long, slender covered with white waxy secretions

Management:

- Triozhophos 2 ml+ neem oil 5 ml/1, phosalone 35 EC 1.5 ml+ neem oil 5 ml/1.
- Release *Cryptolaemous montrouzieri* beetles @ 10/tree.

Spiraling whitefly, *Aleurodicus dispersus*

Symptoms of damage:

- Nymphs and adults suck the sap from leaves
- Honey dew - development of sooty mould fungus
- Yellowing of leaves.
- Dropping of affected leaves.

Identification of pest:

- Nymph - short glass like rods of wax along the sides of the body
- Adult - Powdery white, active during early morning hours.

Management:

- Field sanitation
- Removal of host plants
- Installation of yellow sticky traps
- During heavy infestation - application of imidacloprid 200SL at 0.01%or triazophos 40EC at 0.06%
- Spray neem oil 3% or NSKE 5%
- Release of predators *viz.*, Coccinellid predator, *Cryptolaemus montrouzieri*
- Release of parasitoids *viz.*, *Encarsia haitierrsis* and *E.guadeloupae*

The Jaboticaba

The Jabuticaba (*Plinia cauliflora*) is a fruit-bearing tree in the family Myrtaceae native to Minas Gerais and São Paulo states in southeastern Brazil. Related species in the genus *Myrciaria*, often referred to by the same common name, are native to Brazil, Argentina, Paraguay, and Bolivia. The tree is grown for its purplish-black, white-pulped fruits; they can be eaten raw or be used to make jellies and drinks (plain juice or wine). Other common names include Brazilian Grape Tree, Jaboticaba, Jabotica, Jabuticabeira, Guaperu, Guapuru, Hivapuru, Sabará and Ybapuru (Guarani).

Description

Tree: The tree is a slow-growing evergreen. It has salmon-coloured leaves when they are young, which turn green as they mature. The tree prefers moist, rich, lightly acidic soil. It is widely adaptable, however, and grows satisfactorily even on alkaline beach-sand type soils, so long as they are tended and irrigated. Its flowers are white and grow directly from its trunk in a cauliflorous habit. In an uncultivated state, the tree may flower and fruit only once or twice a year, but when continuously irrigated it flowers frequently, and fresh fruit can be available year round in tropical regions.

Figure: *Jabuticaba tree*

Fruit: The fruit is a thick-skinned berry and typically measures 3–4 cm in diametre. The fruit resembles that of a slip-skin grape. It has a thick, purple, astringent skin that encases a sweet, white or rosy pink gelatinous flesh. Embedded within the flesh are one to four large seeds, which vary in shape depending on the species. Common in Brazilian markets, jabuticabas are largely eaten fresh; their popularity has been likened to that of grapes in the United States. Fresh fruit may begin to ferment 3 to 4 days after harvest, so they are often used to make jams, tarts, strong wines, and liqueurs. Due to the extremely short shelf-life, fresh jabuticaba fruit is very rare in markets outside of areas of cultivation. Traditionally, an astringent decoction of the sun-dried skins has been used as a treatment for hemoptysis, asthma, diarrhoea, and gargled for chronic inflammation of the tonsils.

Several potent antioxidant and anti-inflammatory anti-cancer compounds have been isolated from the fruit. One that is unique to the fruit is jaboticabin.

In Brazil the fruit of several related species, namely *Myrciaria tenella* and *M. trunciflora*, share the same common name. While all jabuticaba species are subtropical and can tolerate mild, brief frosts, some species may be marginally more cold-tolerant. Commercial cultivation of the fruit in the Northern Hemisphere is more restricted by extremely slow growth and the short shelf-life of fruit than by temperature requirements. Grafted plants may bear fruit in 5 years; seed grown trees may take 10 to 20 years to bear fruit, though their slow growth and small size when immature make them popular as bonsai or container ornamental plants in temperate regions. Jabuticabas are fairly adaptable to various kinds of growing conditions, tolerating sand or rich topsoil. They are intolerant of salty soils or salt spray. They are tolerant of mild drought, though fruit production may be reduced, and irrigation will be required in extended or severe droughts.

Cultural Aspects

The name jabuticaba, derived from the Tupi word *Jabuti* (tortoise) + *Caba* (place), meaning the place where you find tortoises. The Guarani name is "Yvapuru", where *yva* means fruit, and the onomatopoeic word *puri* for the crunching sound the fruit produces when bitten.

A traditional song from the eastern region of Bolivia refers to a young woman as having "eyes like the guapuru" (because of their soft blackness) and a mouth "as sweet as the achachairu."

The jabuticaba tree, which appears as a charge on the coat of arms of Contagem, Minas Gerais, Brazil, has become a widely used species in the art of bonsai, particularly in Taiwan and parts of the Caribbean.

In Brazil, it is common to refer to something allegedly unique to the country as a "jabuticaba" since the tree supposedly only grows in Brazil. It is usually a pejorative expression.

Pollination

It has been reported from Brazil that solitary jaboticaba trees bear poorly compared with those planted in groups, which indicates that cross-pollination enhances productivity.

Climate

In Brazil, jaboticabas grow from sea-level to elevations of more than 3,000 ft (910 m). At Minas Gerais, the temperature rarely falls below 33° F (0.56° C). Trees in central Florida have lived through freezing weather. In 1917, one very young jaboticaba tree at Brooksville survived a drop in temperature to 18° F (-7.78° C), only the foliage and branches being killed back. In southern Florida, jaboticabas have not been damaged by brief periods of 26° F (-3.33° C).

Soil

Jaboticaba trees grow best on deep, rich, well-drained soil, but have grown and borne well on sand in central Florida and have been fairly satisfactory in the southern part of the state on oolitic limestone.

Propagation

Jaboticabas are usually grown from seeds in South America. These are nearly always polyembryonic, producing 4 to 6 plants per seed. They germinate in 20 to 40 days. Selected strains can be reproduced by inarching (approach-grafting) or air-layering. Budding is not easily accomplished because of the thinness of the bark and hardness of the wood. Side-veneer grafting is fairly successful. And experimental work has shown that propagation by tissue culture may be feasible.

At the Agricultural Research and Education Centre in Homestead, Florida, 6 related genera, including 10 species, were tried as rootstocks in grafting experiments but none was successful. However, *M. cauliflora* scions were satisfactorily joined to rootstock of the same species 1/8 to 1/4 in (3-6 mm) thick, bound with parafilm and grown in plastic bags under mist.

Culture

Jaboticaba trees in plantations should be spaced at least 30 ft (9 m) apart each way. Dr. Wilson Popenoe wrote that in Brazil they were nearly always planted too close–about 15 ft (4.5 m) apart, greatly restricting normal development.

Growth is so slow that a seedling may take 3 years to reach 18 in (45 cm) in height. However, a seedling tree in sand at Orlando, Florida, was 15 ft (4.5 m) high when 10 years old. Others on limestone at the United States Department of Agriculture's Subtropical Horticulture Research Unit were shrubby and only 5 to 6 ft (1.5-1.8 m) high when 10 and 11 years old. Seedlings may not bear fruit until 8 to 15 years of age, though one seedling selection flowered in 4 to 5 years. Grafted trees have fruited in 7 years. One planted near Bradenton, Florida, in bagasse-enriched soil started bearing the 6th year. The fruit develops quickly, in 1 to 3 months, after flowering.

Traditionally, jaboticabas have not been given fertilizer in Brazil, the belief prevailing that it might be prejudicial rather than beneficial because of the sensitivity of the root system. Some agronomists have advocated digging a series of pits around the base of the tree and filling them with organic matter enriched with 1 part ammonium sulfate, 2 parts superphosphate, and 1 part potassium chlorate. The pits store and gradually release the nutrients and the water from the fall rains.

In 1978, E.A. Ackerman of the Rare Fruit Council International, Inc., reported on fertilizer experiments with 63 one-year-old and 48 two- and three-year-old seedlings in containers. Better growth was obtained with plants in a mixture of equal amounts of acid sandy muck, vermiculite, and peat, given feedings of 32 g of 14-14-14 slow-release fertilizer (Osmocote), roughly every 2 1/2 months, and 3 gallons (11.4 litres) of well water (pH 7.20) by a drip system every 2 days over a period of 18 months, than plants given other treatments. The addition of chelated iron was of no advantage; chelated zinc retarded growth rate, chelated manganese stopped growth and caused defoliation. Abundant water was found to be essential to survival. Irrigation to promote flowering in the dry season is recommended in Brazil to avoid the detrimental effects of flowering in the rainy season.

Season

The time of fruiting varies with the species and/or cultivar and, of course, the locale. In Rio de Janeiro, *M. cauliflora* fruits in May and *M. jaboticaba* in September. If the trees are heavily irrigated in the dry season, they may bear several crops a year. Trees in southern Florida usually produce 2 crops a year.

Harvesting and Packing

In Brazil, jaboticabas harvested in the interior are shipped crudely in second-hand wooden boxes to urban markets. The toughness of the skin prevents serious bruising if the boxes are handled with some care.

Keeping Quality

Jaboticabas, once harvested, ferment quickly at ordinary temperatures.

Pests and Diseases

If the jaboticaba blooms during a period of drought, many flowers desiccate. If blooming occurs during heavy rains, many flowers will be affected by rust caused by a fungus. The variety 'Sabará' is particularly susceptible to attacks of rust on the flowers and fruits. This is the most serious disease of the jaboticaba in Brazil. The initial signs are circular spots, at first yellow then dark-brown.

Fruit-eating birds are very troublesome to jaboticaba growers in Brazil. To protect the crop, double-folded newspaper pages are placed around individual clusters and tied at the top. If birds are very aggressive, or if there are high winds, the paper must be secured with string at the bottom also. To facilitate this operation, it may be necessary in winter or early spring to do some pruning to make it easier to climb the trees and this will result in protecting a larger portion of the crop. Furthermore, reducing the number of fruits has the effect of increasing the size of those that remain. In Florida, raccoons and opossums make raids on jaboticabas.

Lychee

The lychee (*Litchi chinensis*) (Chinese: T*f* g; pinyin: *lì zhī*) is the sole member of the genus *Litchi* in the soapberry family, Sapindaceae. It is a tropical and subtropical fruit tree native to the Guangdong and Fujian provinces of China, and now cultivated in many parts of the world. The fresh fruit has a "delicate, whitish pulp" with a floral smell and a fragrant, sweet flavor. Since this perfume-like flavor is lost in the process of canning, the fruit is usually eaten fresh.

An evergreen tree reaching 10–28 metres tall, the lychee bears fleshy fruits that are up to 5 cm (2.0 in) long and 4 cm (1.6 in) wide. The outside of the fruit is covered by a pink-red, roughly textured rind that is inedible but easily removed to expose a layer of sweet, translucent white flesh. Lychees are eaten in many different dessert dishes, and are especially popular in China, throughout Southeast Asia, along with South Asia and parts of Southern Africa.

The lychee is cultivated in China, Sri Lanka, Taiwan, Thailand, Vietnam, Japan, India, Bangladesh, Pakistan, Nepal, Australia and parts of Southern Africa. China is the main producer, followed by India.

The lychee has a history and cultivation going back to 2000 BC according to records in China. Cultivation began in the area of southern China, Malaysia, and Vietnam. Wild trees still grow in parts of southern China and on Hainan Island. There are many stories of the fruit's use as a delicacy in the Chinese Imperial Court. It was first described and introduced to the West in 1656 by Michael Boym, a Polish Jesuit missionary (at that time Polish–Lithuanian Commonwealth).

Taxonomy

Litchi chinensis was described and named by French naturalist Pierre Sonnerat in his *Voyage aux Indes orientales et à la Chine, fait depuis 1774 jusqu'à 1781* (1782). There are three subspecies, determined by flower arrangement, twig thickness, fruit, and number of stamens.

- *Litchi chinensis* subsp. *chinensis* is the only commercialized lychee. It grows wild in southern China, northern Vietnam, and Cambodia. It has thin twigs, flowers typically have six stamens, fruit are smooth or with protuberances up to 2 mm.
- *Litchi chinensis* subsp. *philippinensis* (Radlk.) Leenh. It is common in the wild in the Philippines and rarely cultivated. It has thin twigs, six to seven stamens, long oval fruit with spiky protuberances up to 3 mm.
- *Litchi chinensis* subsp. *javensis*. It is only known in cultivation, in Malaysia and Indonesia. It has thick twigs, flowers with

seven to eleven stamens in sessile clusters, smooth fruit with protuberances up to 1 mm.

Figure: *Pierre Sonnerat's drawing from* Voyage aux Indes Orientales et à la Chine *(1782)*

Description

Litchi chinensis is an evergreen tree that is frequently less than 19 m (62 ft) tall, sometimes reaching more than 15 m (49 ft). The bark is grey-black, the branches a brownish-red. Leaves are 10 to 25 cm (3.9 to 9.8 in) or longer, with leaflets in 2-4 pairs. Litchee have a similar foliage to the Lauraceae family likely due to convergent evolution. They are adapted by developing leaves that repel water, and are called laurophyll or lauroid leaves. Flowers grow on a terminal inflorescence with many panicles on the current season's growth. The panicles grow in clusters of ten or more, reaching 10 to 40 cm (3.9 to 15.7 in) or longer, holding hundreds of small white, yellow, or green flowers that are distinctively fragrant.

Fruits mature in 80–112 days, depending on climate, location, and cultivar. Fruits reach up to 5 cm (2.0 in) long and 4 cm (1.6 in) wide, varying in shape from round, to ovoid, to heart-shaped. The thin, tough inedible skin is green when immature, ripening to red or pink-red, and is smooth or covered with small sharp protuberances. The skin turns brown and dry when left out after harvesting. The fleshy, edible portion of the fruit is an aril, surrounding one dark brown inedible seed that is 1 to 3.3 cm (0.39 to 1.30 in) long and .6 to 1.2 cm (0.24 to 0.47 in) wide. Some cultivars produce a high percentage of fruits with shriveled aborted seeds known as 'chicken tongues'. These fruit typically have a higher price, due to having more edible flesh.

History

Cultivation of lychee began in the region of southern China, Malaysia, and northern Vietnam. Wild trees still grow in rainforests in Guangdong province and on Hainan Island. Unofficial records in China refer to lychee as far back as 2000 BC.

In the 1st century, fresh lychees were in such demand at the Imperial Court that a special courier service with fast horses would bring the fresh fruit from Guangdong. There was great demand for lychee in the Song Dynasty (960-1279), according to Cai Xiang, in his *Li chi pu* (Treatise on Lychees). It was also the favourite fruit of Emperor Li Longji (Xuanzong)'s favoured concubine Yang Yuhuan (Yang Guifei). The emperor had the fruit delivered at great expense to the capital.

In the Chinese classical work, *Shanglin Fu*, it is related that the alternate name, meaning leaving its branches, is so-called because once the fruit is picked it deteriorates quickly.

The lychee early attracted attention of European travellers. Juan González de Mendoza in his *History of the great and mighty kingdom of China* (1585; English translation 1588), based on the reports of Spanish friars who had visited China in the 1570s, gave the fruit high praise:

They haue a kinde of plummes, that they doo call *lechias*, that are of an exceeding gallant tast, and neuer hurteth any body, although they shoulde eate a great number of them.

The lychee was scientifically described by Pierre Sonnerat (1748–1814) on a return from his travel to China and Southeast Asia. It was then introduced to the Réunion Island in 1764 by Joseph-François Charpentier de Cossigny de Palma. It was later introduced to Madagascar which has become a major producer.

Cultivation of Litchis

Temperature and humidity :

- The average maximum temperature in the litchi-producing areas of South Africa should be at least 23 °C during October and 24 °C during November, with a relative humidity of 50 % and higher.
- The average monthly minimum temperature in areas where litchis are produced should be above 6 °C. Areas where heavy frost occurs are not suitable for litchi production. It should, however, be cold and dry enough in winter to ensure good dormancy.

- The minimum temperature in some Lowveld areas (Malelane and Komatipoort) does not drop low enough in winter to give the trees the proper dormancy period. Trees can be forced into dormancy by withholding water/irrigation during the 3 coldest months of the year. Producers must, however, ensure that especially young trees do not dry out.

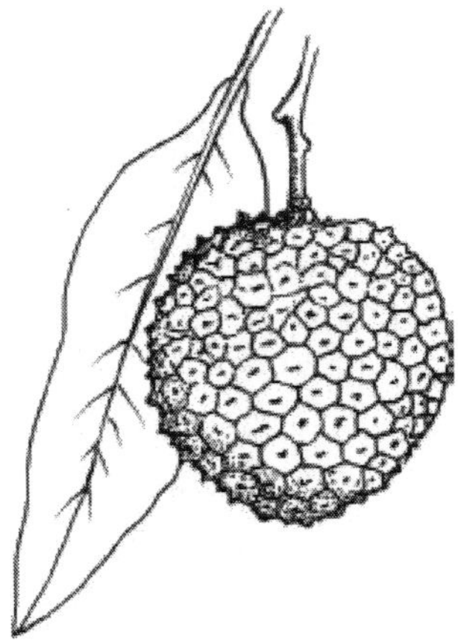

Soil

- Litchis grow very well, especially in sandy soil in the cooler subtropical areas.
- However, the trees also grow and produce well in clay soil in warmer areas.
- Litchis are well adapted to different soil types.

Water Supply

- Because of the varying root distribution in different soils (deep in sandy soils, shallow in clay soils) water is very important for the optimum development of the plant.
- In sandy soils short irrigation cycles with small quantities of water are usually effective.
- In clay soils water is available for longer periods, but it is important that the soil does not become too wet or too dry.

Drainage

- Poorly-drained soil or soil with impenetrable layers shallower than 1 m below the surface is not suitable for litchis.
- Although gravelly or rocky soils drain well, these do not supply enough water to the trees because of poor waterholding capacity. Good irrigation practices, such as wetting the soil more frequently with small quantities of water will make these soils more suitable.

Cultivars

Litchis were originally imported from China, India, Taiwan and Florida, USA. Cultivars grown in South Africa are divided into the following groups:

Mauritius Group

This group is usually planted locally as well as abroad and produces satisfactory yields and fruit of good quality, e.g. H.L.H., Mauritius, Muzaffarpur, Late Large Red, Hazipur, Saharanpur and Rose-Scented.

Chinese Group

These trees produce very poor yields, but the fruit is of excellent quality and has a high percentage of chicken-tongue seeds. Cultivars include Haak Yip, Shang Shou Huai, Kontand, Glutinous Rice and Three Months Red.

Madras Group

These trees bear colourful red fruit, but fruit quality is poor. Cultivars include Kafri, Shorts Seedless, Johnstone's Favourite, Emmerson, Durbhanga, Maries, Mooragusha, Madras 19, Hazipur/Saharanpur, Red McLean, Brewster and Bedana.

Tree Quality

- A good air-layer tree has a single erect stem. The first scaffold branches should branch horizontally at a height of about 200 mm. Any acute forks that branch lower than 200 mm should be avoided.
- In grafted trees the graft-union height should be about 200 mm from the ground so that the first scaffold branches can branch at 300 mm. The graft union must be strongly attached and nurserymen must remove the grafting strip so that girdling cannot occur.

Aftercare of Grafted Trees

Trees can also be propagated by means of grafting.

- Weekly aftercare is very important and suckers and wild shoots that develop on the rootstock below the graft wound must be removed.
- After 5 to 6 weeks the buds start swelling and growing. A small cut can then be made through the plastic next to the bud. The bud grows through this cut, but the plastic strip must not be removed too soon. Once the first new growth has hardened off, the strip can be removed.

- Grafted trees have a better root system than trees developed from air layers and therefore show rapid initial growth. Air layering is, however, preferred to grafting because of a better end product.

Soil Sampling

A representative soil sample should be taken for analysis. A soil sample must represent a homogeneous area where there are no visible soil differences. If there are colour or texture differences the land should be subdivided and separate samples of the different parts should be taken. Use a spade to take the samples.

Soil Preparation

Examine the soil for suitability in respect of depth, drainage and compacted layers. It should preferably be 1 to 2 m deep.

Prepare the soil according to the results of the soil analysis, especially when large quantities of lime are required. If the soil is suitable for litchi production, it must be prepared well in advance.

Before planting, the soil must be tilled as deep and as thoroughly as possible so that it will not be necessary to make the planting holes too big.

If the soil is very acid, heavy lime applications may be necessary. Two-thirds of the recommended quantity of lime must be scattered over the planting area, mixed with the topsoil and then ploughed in as deep as possible, at least 9 to 12 months before planting. Calcium (lime) moves very slowly downwards into the soil and must therefore be worked in to the depth of the root zone.

A cover crop can then be planted and ploughed in about 6 months later to improve the organic matter content of the soil. The remaining lime (one third) and all the required phosphate must be scattered and incorporated at the same time. The trees can be planted 3 months later.

If a lighter lime application (2_4 t/ha) is required, the lime can be worked into the soil at least 3 months before planting and phosphate 1 month before planting.

Planting

Remember that litchi trees have a long life and become large.

Trees should be planted far apart to eliminate competition and to prevent branches of adjoining trees from growing into each other.

The entire outer area of the tree must be exposed to sunlight and air movement.

Planting Distance

A 25-year-old tree can reach a crown diametre of 12 m. If trees are widely spaced and later become uncontrollably big an economic yield will not be possible. If the trees are to be spaced closely together, size must be controlled from the start by pruning. Try to plant as many controllable trees as possible per hectare. The ideal planting distance is 9 x 6 m.

Planting the Trees

Litchi trees can be transplanted any time of the year, but the best time is during spring or at the beginning of the rainy season. Planting holes should be square (in deep-ploughed soil 300 x 500 mm and in nonploughed soil 500 x 500 mm). Mix the topsoil with compost and put it back into the bottom of the hole. When planting the tree, remove the container and loosen the soil around the roots without damaging the roots.

After planting, compress the soil slightly by standing on it. Wet the soil around the tree immediately after planting. Place a mulch around the newly-transplanted tree. Irrigate young trees regularly after planting. They must never suffer from a water shortage or too wet conditions.

Leaf Analysis

Leaf analysis is the only technique according to which sensible fertilisation can be applied to a specific planting. The following aspects are important:

- The correct time for sampling is from mid-September to mid-November.
- The correct leaf must be sampled.
- The first leaf sample of a specific planting must be accompanied by a soil sample.

A leaf and soil sample must represent a planting of not more than 3 ha.

- The sampling method is important:
- Select about 20 healthy trees, well distributed throughout the planting.
- The trees must be of homogeneous appearance and representative of the average trees in the planting.
- Sample 4 leaves per tree.
- Do not take samples from obviously good or weak trees.

Fertilisation

Do not fertilise newly-transplanted trees too soon. Fertiliser should only be applied about 1 year after transplanting. The applications must be very light and broadcast evenly, but not against the stems of the trees. Irrigate after applying fertiliser.

Application and Quantities

- Fertiliser should be broadcast evenly about 0,2 m from the stem to 0,5 m outside the drip area of the tree.
- Irrigate lightly immediately after application. Fertilisers must not be worked into the soil.
- As soon as the trees are established and start growing, fertiliser must be applied regularly according to the quantities given in the table.

Time of Application

Quantity of fertiliser per tree per year according to age (g)

Age years	*LAN 28 % N*	*Superphosphate*	*Potassium chloride*
1	200	250	50
2 - 3	500	250	100
4 - 5	1 000	250	200
6 - 7	1 500	500	300
8 - 9	2 000	500	400
10 - 11	2 500	750	500
12 - 13	3 000	750	750
14 - 15	3 500	1 000	1 000
15 and older	4 000	1 000	1 000

Nitrogen (N)

First year: divide the nitrogen fertiliser into 8 equal monthly applications of 25 g each and apply during summer (September to April).

Second to fifth year: divide the nitrogen fertiliser into 5 equal applications and apply during summer (September to April).

Sixth year and older: half of the nitrogen fertiliser is applied immediately before flowering and the remainder just after harvesting.

Phosphate (P): All the phosphate is applied immediately after harvesting.

Potassium (K): Half of the potassium fertiliser is applied just before flowering and the remainder after harvesting.

Zinc (Zn) and boron (B): Zinc must be applied at least 4 times a year. The following substances and concentrations are recommended per 100 l of water:

- Zinc oxide at 200 g or
- Nitro-Zn at 150 ml or
- Agri-zinc at 50 ml.

Spray the trees from soon after planting with 100 g borax or 75 g Solubor/100 l water every 2 years.

Water Requirements

- Litchi trees need regular watering and therefore it is essential that enough water must be available from the flowering stage until after the February/March flush following the harvest.
- Because the edible portion of the litchi fruit has a water content of 86 %, the availability of water remains important during the development period.
- A water shortage will delay development of the fruit and adversely affect the size, mass and quality of the litchis.
- Irrigation must continue after harvesting to ensure that a normal growth flush occurs during February/March, just before the beginning of the dormant period.
- During dormancy (April to July) irrigation should be reduced, but the tree should not suffer drought.
- Young trees that are not producing yet are irrigated throughout the year.
- Producers normally stop irrigating the trees during the coldest months of the year (June and July) so that they can have a proper dormant period. In areas where it is never very cold, irrigation should stop to force the trees into dormancy.

Covering Litchi Fruit Clusters

- Covering the fruit with paperbags prevents damage from fruitflies and litchi moths, as well as sunburn and cracking.
- The best time for covering the clusters will vary from one locality to the next.
- Paperbags are at present the most suitable and also the cheapest covering material.
- The fruit develops a very attractive red colour inside the paperbags.

- Covering the fruit also extends the harvest period because fruit can be left on the tree for a longer period.
- The paperbag at present used is open at both ends. It is made from reinforced brown paper that is very resistant to the elements.
- The fruit must, however, be covered in the correct way and during the right time of the day (not early in the morning or immediately after rain), because of the risk of decay when covered while still wet. Spray against litchi moth and fruitflies just before covering.
- Remove all the leaves on the cluster stem before covering.
- Do not put more than 25 litchis in one bag.
- The bags must be fixed to hang to an angle to allow water to flow out.
    - o Both sides of the bag must be closed and, if necessary, a small opening must be left for rainwater to run out. The open ends can easily and quickly be stapled together.
    - o The paperbags will even keep fruit bats and birds away.
    - o Another advantage is that the fruit is handled less during harvesting and this limits damage caused by rough handling.
    - o When removed carefully, the bags can be used for another season.

Insect Pests

- Important pests include bark borers, litchi moth and fruitflies.
- Consult your nearest extension officer about ways to control these insects.

Harvesting

- The stage of maturity at which fruit is harvested is one of the most important factors that determine the ultimate quality at the point of sale.
- Litchis do not develop further after picking. The fruit must therefore remain on the tree until quite ripe.
- Litchis harvested too early have an unattractive colour and have a sour taste.

4

Temperate Fruit Crops

Berry

The botanical definition of a berry is a fleshy fruit produced from a single ovary. Grapes and avocados are two common examples. The berry is the most common type of fleshy fruit in which the entire ovary wall ripens into an edible pericarp. They may have one or more carpels. The seeds are usually embedded in the fleshy interior of the ovary, but there are some non-fleshy exceptions, such as peppers, that have space rather than pulp around their seeds.

A plant that bears berries is said to be bacciferous or baccate (a fruit that resembles a berry, whether it actually is a berry or not, can also be called "baccate").

In everyday English, "berry" is a term for any small edible fruit. These "berries" are usually juicy, round, brightly coloured, sweet or sour, and do not have a stone or pit, although many seeds may be present.

Many berries, such as the tomato, are edible, but others in the same family, such as the fruits of the deadly nightshade (*Atropa belladonna*) and the fruits of the potato (*Solanum tuberosum*) are poisonous to humans.

Botanical Berries

In botanical language, a berry is a simple fruit having seeds and pulp produced from a single ovary; the ovary can be inferior or superior.

Figure: *Ripe lingonberries*

Modified berries

The fruit of citrus, such as the orange, kumquat and lemon, is a berry with a thick rind and a very juicy interior that is given the special name hesperidium.

Berries which develop from an inferior ovary are sometimes termed epigynous berries or false berries, as opposed to true berries which develop from a superior ovary. In epigynous berries, the berry includes tissue derived from parts of the flower besides the ovary. The floral tube, formed from the basal part of the sepals, petals and stamens can become fleshy at maturity and is united with the ovary to form the fruit. Common fruits that are sometimes classified as epigynous berries include bananas, coffee, members of the genus *Vaccinium* (e.g., cranberries and blueberries), and members of the family Cucurbitaceae (e.g., cucumbers, melons and squash).

Another specialized term is also used for Cucurbitaceae fruits, which are modified to have a hard outer rind, and are given the special name pepo. While pepos are most common in the Cucurbitaceae, the fruits of *Passiflora* and *Carica* are sometimes also considered pepos.

Fruits not Botanical Berries

Many fruits commonly referred to as berries are not actual berries by the scientific definition, but fall into one of the following categories:

Drupes: Drupes are fleshy fruits produced from a (usually) single-seeded ovary with a hard stony layer (called the endocarp) surrounding the seed. Familiar examples include the stonefruits of *Prunus* species (peaches, plums and cherries), olives, coconut and bayberry. Other drupe-like fruits with a single seed that lack the stony endocarp include sea-buckthorn (*Hippophae rhamnoides*, Elaeagnaceae), an achene surrounded by the swollen hypanthium, which provides the fleshy layer.

Pomes: The pome fruits produced by plants in subtribe Pyrinae of family Rosaceae, such as apples and pears, have a structure (the core) that clearly separates the seeds from the ovary tissue. However, some of the smaller pomes are sometimes referred to as berries. Bright red haws from *Crataegus* are sometimes called hawberries. *Amelanchier* pomes become so soft at maturity that they resemble a blueberry and are known as Juneberries or Saskatoon berries.

Aggregate Fruits: Aggregate fruits contain seeds from different ovaries of a single flower. Examples include blackberry and raspberry.

Figure: *Alaska wild "berries" from the Innoko National Wildlife Refuge, a mixture of true berries and aggregate fruits*

Multiple fruits: Multiple fruits include the fruits of multiple flowers that are merged or packed closely together. The mulberry is a berry-like example of a multiple fruit; it develops from a cluster of tiny separate flowers that become compressed as they develop into fruit.

Accessory Fruits: In accessory fruits, the edible part is not generated by the ovary. Berry-like examples include:

- Strawberry - the aggregate of seed-like achenes is actually the "fruit", derived from an aggregate of ovaries, and the fleshy part develops from the receptacle.
- Gurbir, *Duchesnea indica* - structured just like a strawberry
- Sea grape (*Coccoloba uvifera*; Polygonaceae) - the fruit is a dry capsule surrounded by fleshy calyx
- Wintergreen (*Gaultheria procumbens*) - the fruit is a dry capsule surrounded by fleshy calyx

Colour and Potential Health Benefits

Berries are typically of a contrasting colour to their background (often of green leaves), making them visible and attractive to frugivorous animals and birds. This assists the wide dispersal of the plants' seeds.

Berry colours are due to natural plant pigments, many of which are polyphenols, such as the flavonoids, anthocyanins, and tannins, localized mainly in berry skins and seeds. Berry pigments are usually antioxidants *in vitro*. However, there is no physiological evidence established to date that berry polyphenols have actual antioxidant value within the human body, and it is not permitted to claim that polyphenols have antioxidant health value on product labels in the United States and Europe.

Blueberry

Blueberries are perennial flowering plants with indigo-coloured berries from the section *Cyanococcus* within the genus *Vaccinium* (a genus that also includes cranberries and bilberries). Species in the section *Cyanococcus* are the most common fruits sold as "blueberries" and are native to North America (commercially cultivated highbush blueberries were not introduced into Europe until the 1930s).

Blueberries are usually erect. Prostrate shrubs can vary in size from 10 centimetres (3.9 in) to 4 metres (13 ft) in height. In the commercial production of blueberries, the smaller species are known as "low-bush blueberries" (synonymous with "wild"), while the larger species are known as "high-bush blueberries".

The leaves can be either deciduous or evergreen, ovate to lanceolate, and 1–8 cm (0.39–3.15 in) long and 0.5–3.5 cm (0.20–1.38 in) broad. The flowers are bell-shaped, white, pale pink or red, sometimes tinged greenish. The fruit is a berry 5–16 millimetres (0.20–0.63 in) in diametre with a flared crown at the end; they are pale greenish at first, then reddish-purple, and finally dark purple when ripe. They are covered in a protective coating of powdery epicuticular wax, colloquially known as the "bloom". They have a sweet taste when mature, with variable acidity. Blueberry bushes typically bear fruit in the middle of the growing season: fruiting times are affected by local conditions such as altitude and latitude, so the peak of the crop can vary from May to August depending upon these conditions.

Cultivation

Blueberries may be cultivated, or they may be picked from semiwild or wild bushes. In North America, the most common cultivated species is *V. corymbosum*, the northern highbush blueberry. Hybrids of this with other *Vaccinium* species adapted to southern U.S. climates are known collectively as southern highbush blueberries.

So-called "wild" (lowbush) blueberries, smaller than cultivated highbush ones, are prized for their intense colour. The lowbush blueberry,

V. angustifolium, is found from the Atlantic provinces westward to Quebec and southward to Michigan and West Virginia. In some areas, it produces natural "blueberry barrens", where it is the dominant species covering large areas. Several First Nations communities in Ontario are involved in harvesting wild blueberries.

Figure: *Blueberry flowers*

"Wild" has been adopted as a marketing term for harvests of managed native stands of lowbush blueberries. The bushes are not planted or genetically manipulated, but they are pruned or burned over every two years, and pests are "managed".

Numerous highbush cultivars of blueberries are available, with diversity among them, each having a unique flavor. The most important blueberry breeding programme has been the USDA-ARS breeding programme based at Beltsville, Maryland, and Chatsworth, New Jersey. This programme began when Frederick Coville of the USDA-ARS collaborated with Elizabeth Coleman White of New Jersey. In the early part of the 20th century, White offered pineland residents cash for wild blueberry plants with unusually large fruit.

The rabbiteye blueberry (*Vaccinium virgatum* syn. *V. ashei*) is a southern type of blueberry produced from the Carolinas to the Gulf Coast states. Other important species in North America include *V. pallidum*, the hillside or dryland blueberry. It is native to the eastern U.S., and common in the Appalachians and the Piedmont of the Southeast. Sparkleberry, *V. arboreum*, is a common wild species on sandy soils in the Southeast.

Growing Areas

Significant production of highbush blueberries occurs in British Columbia, Maryland, Western Oregon, Michigan, New Jersey, North Carolina, and Washington. The production of southern highbush varieties in California is rapidly increasing, as varieties originating from University of Florida, Connecticut, New Hampshire, North Carolina State University and Maine have been introduced. Southern highbush berries are now also cultivated in the Mediterranean regions of Europe, Southern Hemisphere countries and China.

Harvesting

The blueberry harvest in North America varies. It can start as early as May and usually ends in late summer. The principal areas of production in the Southern Hemisphere (Australia, Chile, New Zealand and Argentina) have long periods of harvest. In Australia, for example, due to the geographic spread of blueberry farms and the development of new cultivation techniques, the industry is able to provide fresh blueberries for 10 months of the year – from July through to April. Similar to other fruits and vegetables, climate-controlled storage allows growers to preserve picked blueberries. Harvest in the UK is from June to August.

For many years, blueberries were hand picked. In modern times, traditional hand picking is still quite common especially for the more delicate varieties. More commonly, farmers will use harvesters that will shake the fruit off the bush. The fruit is then brought to a cleaning/packaging facility where it is cleaned, packaged, then sold.

Uses

Blueberries are sold fresh or processed as individually quick frozen (IQF) fruit, purée, juice, or dried or infused berries, which in turn may be used in a variety of consumer goods, such as jellies, jams, blueberry pies, muffins, snack foods and cereals.

Blueberry jam is made from blueberries, sugar, water, and fruit pectin.

Blueberry wine is made from the flesh and skin of the berry, which is fermented and then matured; usually the lowbush variety is used.

Nutrients, Phytochemicals and Research

Blueberries have a diverse range of micronutrients, with moderate levels (relative to respective Dietary Reference Intakes) of the essential dietary mineral manganese, vitamin C, vitamin K and dietary fibre (table). One serving provides a relatively low glycemic load score of 4 out of 100 per day.

Blueberries contain anthocyanins, other pigments and various phytochemicals, which are under preliminary research for their potential role in reducing risks of diseases such as inflammation and cancer. Similar to red grape, blueberries may contain resveratrol.

Most polyphenol studies have been conducted using the highbush cultivar of blueberries (*V. corymbosum*), while content of polyphenols and anthocyanins in lowbush (wild) blueberries (*V. angustifolium*) exceeds values found in highbush cultivars.

In preliminary research, feeding blueberries to rats reduced brain damage in experimental stroke and may cause increased production of vascular nitric oxide that influences blood pressure regulation. Additional research showed that blueberry consumption in rats altered glycosaminoglycans that are vascular cell components affecting control of blood pressure.

Other preliminary studies found blueberry consumption lowered cholesterol and total blood lipid levels, possibly affecting symptoms of heart disease. Wild South American varieties may contain higher levels of polyphenols, but this finding remains unconfirmed and uncertain of its significance.

Other preliminary research showed that supplementation of diets with wild blueberry juice may benefit the brain, improve memory and learning in older adults, while possibly reducing blood sugar and symptoms of depression.

Pesticides

The application of pesticides is common in large-scale blueberry monoculture in Maine. Because "wild" is a marketing term generally used for all low-bush blueberries, it is not an indication that such blueberries are free from pesticides.

The Environmental Working Group, referencing the USDA, rates blueberries as a "significant concern".

Pest Management

Blueberries have few serious insect pests and diseases. Gardeners who use Integrated Pest Management (IPM) practices may never have a need to apply pesticides to their plants. IPM is a sustainable approach that allows gardeners to reduce pests to a tolerable level by using the best balance of cultural, physical, biological, & chemical management strategies. IPM takes into account the level of damage a pest is capable of causing, as well as the possible risks to humans and the environment associated with each pest management strategy.

In order for IPM to be effective, home gardeners must be able to recognize common blueberry pests and the damage they cause.

Cultural controls can be quite effective in protecting blueberries from most insect pests and diseases. Good cultural practices such as soil preparation, pruning and cultivar selection all contribute to healthy blueberry plants.

In IPM, pesticide sprays are used only when cultural controls are not effective or as a supplement to cultural controls. If using pesticides, gardeners should choose an effective product that has the lowest possible impact on human health, non-target organisms like bees and the environment. Information on the correct way to apply specific pesticides can be found on the product label. If pesticides are necessary, always use them exactly as directed by the product label as mandated by federal law.

Preventing Pests before and During the Establishment Year

The key to growing healthy blueberries is to plant in the correct soil type or to properly amend the soil before planting. Blueberry plants can live up to 50 years, and healthy blueberry plants are easy to maintain after they have been established. Blueberries need more soil and site preparation than most other fruit. Always get a soil test when choosing a site and amend the soil according to the soil test.

Blueberry plants prefer a sandy, acidic soil, with a pH less than 5.5. They will grow better in all sites if acidic peat is added to the planting soil. Planting blueberries in sheltered areas helps to protect plants from winter injury. Choose a site that is sheltered but receives full sun as too much shade will decrease blossom production and yield.

Weeds

Weeds can stunt blueberry plants and reduce yields. Perennial weeds that can choke out blueberry plants include quackgrass, Canada thistle and raspberries. If quackgrass or Canada thistle are growing in the site to be planted to blueberries, kill the weeds the summer before planting. Both quackgrass and thistles are hard to kill by hoeing and pulling. These weeds are extremely difficult to remove from an established blueberry planting. The easiest way to kill quackgrass is to spray the area with glyphosate in late fall. The best way to kill thistle is to spray with glyphosate or clopyralid when the weeds are forming flower buds in late June. Avoid planting blueberries next to a raspberry patch. Raspberries spread through underground runners or through tipping and can easily crowd out blueberry bushes. Remove any wild brambles from the area before planting.

Annual weeds need to be controlled after planting. During the planting year, pre-emergent herbicides can harm blueberry plants. By using a combination of hand weeding and a woodchip or pine needle mulch, gardeners can avoid the potential harm from herbicides while improving the soil for the blueberries. Wood chip mulches and pine needles slowly decay, releasing nutrients into the soil and should be augmented once a year. Sawdust mulches are not recommended, because sawdust can tie up soil nitrogen, causing the plants to be stunted. Landscape fabric can be used to reduce weed problems in the first few years. Some hardy blueberry cultivars can spread underground through rootsuckers. The landscape fabric will prevent the blueberries from spreading by root suckers.

Once a blueberry patch is established, few weeds sprout through the mulch, and those weeds that do sprout are easy to remove by hand. Trees and shrubs can be a problem in established blueberry fields. Birds can deposit seeds from a variety of trees and shrubs including chokecherry, honeysuckle, raspberry and elderberry. Always remove tree seedlings before the root systems become too large. The best time to kill or remove tree seedlings is from the middle of summer to late fall. By the second year, the trees are often too large to pull.

Blackberry and Raspberry

Blackberries and Raspberries, often termed "Brambles", are a diverse group of species and hybrids in the genus *Rubus*. They are members of the *Rosaceae* family, closely related to strawberry in the subfamily *Rosoideae*. *Rubus* is one of the most diverse genera of flowering plants in the world, consisting of 12 subgenera, some with hundreds of species.

Some of the most important commercially grown brambles are actually blackberry – red raspberry hybrids. Examples include 'Boysenberry', 'Loganberry', and 'Youngberry'. The fruit flavor is unique, but culture and management is more like blackberry than raspberry. Major *Rubus* hybrids include:

Cultivars

Blackberries. Cultivars are classified by growth habit – trailing, erect, semi-erect, and also as thorny or thornless. In the Pacific Northwest, trailing types comprise 98% of acreage, like 'Thornless Evergreen', 'Marion', and 'Kotata'. Marion is 50% of the acreage; growers in Oregon call them "Marionberries", not blackberries. 'Thornless Evergreen' is a spine-free mutant of the European species *R. laciniatus*.

Red Raspberries. As with blackberry, cultivars are distinguished by growth habit, trailing or erect. Primocane or autumn fruiting raspberries are a unique group that can produce fruit in their first year, or two crops per year. 'Heritage', 'Amity', 'Redwing', and 'Autumn Bliss' are major primocane fruiting types. In the Pacific Northwest, 'Meeker' is grown on about 60% of the acreage, 'Willamette' on 30%, and a few others are grown on the remaining 10%.

Black Raspberries. Few cultivars of this species are grown. 'Munger' is grown on >90% of the acreage in Oregon, and 'Allen', 'Blackhawk', 'Bristol', and 'Jewel' are available for home garden culture.

Origin of Rubus, History of Cultivation

Blackberry. Blackberries are native to Asia, Europe, North and South America. However, blackberries grown in specific regions are largely derived from species indigenous to that region. Blackberries have been used in Europe for over 2000 years, for eating, medicinal purposes, and as hedges to keep out marauders.

Red Raspberry. The species is indigenous to Asia Minor and North America although the epithet denotes Mt. Ida, in the Caucasus mountains of Asia Minor. Fruits were gathered from the wild by the people of Troy and the foothills of Mt. Ida around the time of Christ. Records of domestication were found in 4th century writings of Palladius, a Roman agriculturist, and seeds have been discovered at Roman forts in Britain. Hence, the Romans probably spread cultivation throughout Europe.

Black Raspberry. This species is indigenous only to North America, where it is most abundant in the east, exclusive of the Gulf states, and found in the west along with related *R. leucodermis*. Domestication appears to have been delayed until the 1800s, due to the popularity of red raspberries and abundant supply of wild fruit.

Botanical Description

Plant: Blackberries and raspberries are erect, semi-erect, or trailing, generally thorny shrubs, producing renewal shoots from the ground called "canes". The plants are perennial, composed of biennial canes which overlap in age. Individual canes grow vegetatively for one year, initiate flower buds in late summer, fruit the following summer, then die. The first year canes are called "primocanes", and in the second year when they flower, "floricanes". Raspberry thorns are finer and more flexible than thorns on blackberries. Within raspberries, black and purple have more prominent thorns than red raspberries. Leaves are palmately compound with 3-5 leaflets, the middle one being the largest. Leaf margins are finely serrate.

Red raspberries are unique in that there are some primocane (syn. Autumn fruiting) cultivars available which produce fruit in their first summer, and allow the crop to be treated more-or-less as an annual.

Flower: White to pink flowers (13 diametre) are borne terminally on several-flowered racemes, cymes, or corymbose inflorescences on current season's growth. For dewberry, raspberries, and some wild blackberries, inflorescences are cymose, and some flowers are borne singly in axils of leaves on fruiting laterals. Blackberry flowers generally have larger petals than those of raspberries. Flowers are initiated in late summer in biennial types, early to mid-summer in primocane fruiting types. The gynoecium consists of 60-100 ovaries, each of which develops into a drupelet. There are 60-90 stamens, five sepals and five petals.

Pollination: Most cultivars of blackberries, black raspberries, and raspberries are self-fruitful and do not require pollinizers. However, dewberries are self-incompatible, and must be inter-planted for good fruit set. Honey bees are naturally attracted to brambles, and wind also aids pollination.

Fruit: In all brambles, the fruit is an aggregate of drupelets. Blackberries retain the receptacle within the fruit at harvest. Fruiting begins in the second year of the planting, and continues for >10 years if properly managed. Fruit development occurs rapidly, taking only 30-50 days for most raspberries, and 40-70 days for blackberries.

General Culture

Soils and Climate: Brambles are suited to a wide range of soil types, from sandy to clay loams, provided with good drainage and pH of 5-7. Depth is not important since plants develop shallow root systems, although blackberries are more deeply rooted and therefore drought tolerant than raspberries as a rule.

Soils previously planted to solanaceous crops (pepper, tomato, eggplant, potato) in the last 5 years should be avoided, since a fungus that can kill raspberry roots is harbored by these plants.

Climatically, blackberries have a more warm-temperate adaptation than do raspberries. Black and red raspberries generally require cooler summers than blackberries, and are poorly adapted to the southern USA or to hot, arid climates. Mid-winter cold hardiness can be a great limitation in growing blackberries in the northern USA or away from the moderate, maritime climates of Oregon and Washington. Tender cultivars can be killed at 0 to +10 F, and the hardiest tolerate only about -10 F. Raspberries are more cold hardy than blackberries, with red hardier than black or purple raspberry.

Red raspberry can tolerate -20°F, whereas black and purple raspberries are injured at -5 to -10°F. Frost damage is generally not a problem for brambles. Blackberries generally have lower chilling requirements to break dormancy than raspberries, ranging from 200-800 hr. Raspberries have chilling requirements ranging from 800-1600 hr.

Propagation: Unlike many fruit crops, brambles are easily propagated by a number of techniques. In fact, most propagate themselves quite well, and can be invasive or even weedy.

Root Cuttings. Roots about pencil-size are cut into pieces 4-6 inches long and planted 2-3 inches deep, 1-2 ft apart in rows. Root cuttings are dug in Jan – Feb., and planted in Mar-April. There are notable exceptions to root cutting propagation. Trailing blackberries, as well as black and purple raspberries, do not produce adventitious buds on roots, (or do not produce enough) and cannot be propagated this way. They are tip-layered. Thornless mutants of trailing cultivars ('Thornless Evergreen', 'Thornless Logan', and 'Thornless Youngberry') are periclinal chimeras, and will produce thorny shoots if propagated by root cuttings.

Suckering. Basically the same as root cuttings, except the adventitious buds have already produced canes.

Tip Layering. This is used for black raspberries, hybrids, trailing types, and thornless mutants in lieu of root cuttings. In late summer, when canes have numerous shoot tips, shoot tips are buried in shallow holes. Lateral roots form on stems, and the following spring, layered plants can be dug and transplanted.

Leafy Stem Cuttings. This method can be used for any bramble; semi-hardwood cuttings are rooted under mist in mid-summer.

Tissue Culture. This is used to a limited extent, especially when virus-free material is difficult to obtain by other means.

Rootstocks – None

Planting Design, Training, Pruning

Most erect-growing Rubus are grown in hedgerows, spaced 8-16 ft apart (commonly 12 ft) to allow for equipment movement. Hedges are kept to heights of 4-6 ft, widths of 30 inches on top, with canes arising from a 123 strip down the row to facilitate mechanical harvesting equipment. Trailing types must be trained on a trellis, usually 2 or 3 wires spaced 18 inches apart, with the top at 5-6 ft. Primocanes are trained by spiralling them around the wires, spacing them as evenly as possible. Alternatively, they may be bundled and tied to the trellis at regular intervals.

Alternate year production is one method of simplifying bramble training. Canes are mowed to the ground every-other year, and produce a crop in alternate years. Costs of production are substantially reduced.

Primocane raspberry culture – As mentioned above, primocane or "Autumn fruiting" raspberries produce fruit terminally on current season's canes. This allows growers to prune all canes to the ground each year without a yield penalty, greatly reducing disease and insect problems, and simplifying cultural practices. All that is needed is a simple support system, as canes top-heavy with fruit tend to bend over.

For erect cultivars that are not mowed for alternate year production, three main pruning practices are required each year: 1) topping, 2) dead floricane removal, and 3) primocane thinning. The basic pruning needsof trailing and some semi-erect cultivars differ only in the lack of need for cane topping. Primocanes can be shortened to 6-82 during the winter when floricane removal and cane thinning are practiced.

The Apple

The apple is the pomaceous fruit of the apple tree, *Malus domestica* of the rose family (Rosaceae). It is one of the most widely cultivated tree fruits, and the most widely known of the many members of genus *Malus* that are used by humans. Apples grow on deciduous trees which are large if grown from seed, but small if grafted onto roots (rootstock). The tree originated in Central Asia, where its wild ancestor, *Malus sieversii*, is still found today. Apples have been grown for thousands of years in Asia and Europe, and were brought to North America by European colonists. Apples have been present in the mythology and religions of many cultures, including Norse, Greek and Christian traditions. In 2010, the fruit's genome was decoded as part of research on disease control and selective breeding in apple production.

There are more than 7,500 known cultivars of apples, resulting in a range of desired characteristics. Different cultivars are bred for various tastes and uses, including cooking, fresh eating and cider production. Domestic apples are generally propagated by grafting, although wild apples grow readily from seed. Trees are prone to a number of fungal, bacterial and pest problems, which can be controlled by a number of organic and non-organic means.

About 69 million tons of apples were grown worldwide in 2010, and China produced almost half of this total. The United States is the second-leading producer, with more than 6% of world production. Turkey is third, followed by Italy, India and Poland. Apples are often eaten raw, but can also be found in many prepared foods (especially desserts) and drinks. Many beneficial health effects are thought to result from eating apples; however, two forms of allergies are seen to various proteins found in the fruit.

About 63 million tonnes of apples were grown worldwide in 2012, with China producing almost half of this total. The United States is the second-leading producer, with more than 6% of world production. The largest exporters of apples in 2009 were China, U.S., Turkey, Poland, Italy, Iran, and India while the biggest importers in the same year were Russia, Germany, the UK and the Netherlands.

***Figure:** Blossoms, fruits, and leaves of the apple tree (Malus domestica)*

In the United States, more than 60% of all the apples sold commercially are grown in Washington. Imported apples from New Zealand and other more temperate areas are competing with U.S. production and increasing each year.

Most of Australia's apple production is for domestic consumption. Imports from New Zealand have been disallowed under quarantine regulations for fireblight since 1921.

Other countries with a significant production are Brazil, Argentina, Ukraine, Germany and South Africa.

Integrated Pest Management

There are a large number of insect and disease pests that attack apples and pears. Good pest management begins with good horticultural management. Removing last year's leaves, or at least mowing them, helps reduce the population of overwintered apple scab fungus spores and some insect pests. Opening the tree up to sunlight and air movement not only improves spray coverage, it reduces humidity for scab and other fungal diseases and promotes fruit bud formation and overall tree health. Weed control is important to prevent competition with newly planted trees. Once trees are established, managing ground cover vegetation in a home orchard by mowing is adequate. Maintaining bare soil around the base of an apple tree is not recommended as the presence of ground cover provides insulation against low winter soil temperatures.

Whether and what to spray apple trees for disease and insect pests depends on your objectives, i.e. what percentage unblemished fruit you desire. If the trees are more than 10 feet tall, spraying becomes physically difficult and time consuming.

Apple scab is the most common disease of apples. Unless you know that an apple tree is a scab-resistant variety, then it's very likely to be scab susceptible. If the tree was unprotected last year and did not have brown spots, dead leaves, and no rough corky spots on the apples, then the cultivar is either scab-resistant or blessed with unusually low scab inoculum in the area.

A high level of scab control on susceptible varieties requires beginning protection when buds open and green tissue emerges (date depends on location, late April to early May in most of Maine) and renewing protection about every 7 days or 2 inches of rain until about two weeks after the flower petals fall. In central Maine, two weeks after petal fall is in mid-June.

The multipurpose home fruit tree spray products are a good choice for backyard apple trees because they simplify the mixing of spray solution which can get a little complicated when dealing with separate fungicides and insecticide products purchased in concentrate form. Also the package size is appropriate. While most products can be stored over the winter if kept unfrozen, cool and dry (for powder formulations), it is best to only buy what will be used this growing season and start fresh each year.

The limitation of the multipurpose products is that because they contain an insecticide plus fungicide, spray should not be applied during bloom when honeybees and other pollinators are visiting the tree. Applications made after twilight when bee activity stops for the day are less damaging, but it is better to avoid spraying multipurpose product altogether while there are open blossoms on the tree. Applying multipurpose spray up until bloom, with a spray just before blossoms open, and then coming back with a spray just after the blossoms fall can provide adequate control for a backyard apple tree if perfection is not the standard.

Sulfur is sold as a fungicide for apple trees, and some sulfur products are approved for organic certification. Sulfur requires more frequent application than home fruit tree spray mixes containing captan, a conventional fungicide that is more reliable for disease control than sulfur.

There are too many potential insect pests to name them all here. A major pest of young home apple trees is the roundheaded apple tree borer. Trees are less likely to be killed by borers once they have developed substantial trunks (over 4 inch trunk diametre).

The main fruit-damaging insects are plum curculio, codling moth, and apple maggot. Applying a home fruit spray product that contains an insecticide when 95+% of the apple blossoms fall off the tree and then again about 10 days later will reduce damage by plum curculio and codling moth. These two sprays also reduce damage from European apple sawfly, leafrollers, and roundheaded apple tree borer. For borer control, in addition to spraying the foliage, some spray should also be directed at the trunk.

In most situations and years, another application of a multi-purpose spray product is not needed until mid-July to protect against apple maggot. One of two more applications at two week intervals starting (i.e. repeat sprays on July 30 and August 15) are usually enough to keep second generation codling moth and apple maggot from being major problems. A final spray in mid-late August may not be necessary for early harvested varieties, but is important for later-harvested cultivars (i.e. those that ripen in late September and October). The late spray on late cultivars also reduces development of sooty blotch and flyspeck fungi that stain fruit.

Insect control using only organic options is more complex and expensive. Surround is a highly formulated clay powder product you can apply to repel many insect pests. Ordinary clay will not be effective. The desired results can be achieved, but you have to keep the trees literally white with the clay powder from petal fall until early July to get good results against plum curculio and codling moth. An alternative is to use several applications of Pyganic (pyrethrin) after petal fall to reduce plum curculio and a couple applications of a Bt product to reduce codling moth.

You can also repel apple maggot flies that begin attacking apples in July with Surround, but by spraying that late in the summer you will still have the clay power residue on the fruit at harvest. Bt does not work against apple maggot, and as a pyrethrin insecticide, Pyganic breaks down too quickly to give prolonged control against apple maggot. Entrust is an effective organic insecticide for apple maggot, but it costs hundreds of dollars for a one pound package that is larger than what most home orchards will need for many years. The active ingredient in Entrust is spinosad and is found in some home spray products, but those product may have other ingredients that disqualify them for organic certification.

You can reduce apple maggot damage by hanging sticky traps at 1 trap per bushel of apples (i.e. one trap per tree for small trees, up to 4 or 5 traps per tree for larger trees).

The main problem that leads to pest control failures when people have sprayed their trees with an appropriate product at the right times is not applying enough water to deliver the correct amount pesticide. The mixing instructions on multipurpose fruit tree pesticide products are based on spraying the tree to run-off, i.e. until the leaves are saturated with water and applying more water would just result in run-off. This is called "dilute" spray. The idea is to apply enough water to almost, but not quite, get to the "run-off" point when spraying more water would just run off the tree. If the spray is just misted on, there won't be enough of the fungicide and insecticide to get desired protection.

Water is heavy to carry, and refilling takes time. To expedite spraying one can spray half, a third, or a fourth as much of that amount of water and deliver the correct amount of pesticide by increasing the amount per gallon by 2X, 3X, or 4X to compensate for the reduced spray volume. The best way to figure out how much water is needed to reach run-off is by spraying with plain water (no pesticide) and seeing how much water it takes.

Pejibaye (Peach Palm)

Bactris gasipaes is a species of palm native to the tropical forests of South and Central America.

Its English language common names include peach-palm. Names from Spanish-speaking countries include *pejibaye* (Costa Rica), *chontaduro* (Colombia, Ecuador), *pijuayo* (Peru), *pijiguao* (Venezuela), *tembé* (Bolivia), and *pibá* (Panama). In Brazilian Portuguese it is called *pupunha*. In Trinidad and Tobago it is known as *peewah*.

Bactris gasipaes, like most sea-island palms, grows erect, with a single slender stem or, more often, several stems to 8 in (20 cm) thick, in a cluster; generally armed with stiff, black spines in circular rows from the base to the summit. There are occasional specimens with only a few spines. It can typically grow to 20 metres (66 ft) or taller . The leaves are pinnate, 3 metres (9.8 ft) long on a 1 metre (3.3 ft) long petiole. The fruit is a drupe with edible pulp surrounding the single seed, 4–6 cm long and 3–5 cm broad. The rind (epicarp) of the fruit can be red, yellow, or orange when the fruit is ripe, depending on the variety of the palm.

Ecology

The fruit is food for many animals, including parrots of the family Psittacidae.

Cultivation

Bactris gasipaes has been used for food for centuries. Spanish explorers found a pejibaye plantation of 30,000 trees on the Atlantic coast of Costa Rica, providing fruit that replaced corn in the indigenous diet. The fruit is stewed in salted water and peeled, the seed is removed, and it may be flavored with salt or honey. The texture both raw and cooked has been compared to a firm sweet potato, and the flavor to hominy or dry squash. The fruit halves may be filled with mayonnaise or sour cream. Raw pejibaye contains irritating acid crystals, so it is often preferred cooked. The raw fruit spoils quickly but it can be stored as a dry meal or preserves. It can yield flour and edible oil.

This plant may also be harvested for heart of palm, and has commercial advantages in being fast growing; the first harvest can be from 18 to 24 months after planting. Brazil has a large domestic market for heart of palm and international demand is growing. It is also an economically important crop in Costa Rica. It is a viable substitute for other sources of heart of palm, such as overexploited native species of *Euterpe*, including *Euterpe oleracea* (*açaí*) and *Euterpe edulis* (*juçara*). It could also become a replacement crop for the threatened Fiji sago palm (*Metroxylon vitiense*).

Pests and Diseases

The trunk of the tree can be infested with *Phytophthora* water molds. The foliage is infested with fungi of the genera *Pestalotiopsis*,

Mycosphaerella, and *Colletotrichum*. The fruit is attacked by fungi of the genera *Monilinia* and *Ceratocystis*. Other pests include mites and insects such as the sugar cane weevil (*Metamasius hemipterus*).

Lemon

The lemon (*Citrus* × *limon*) is a small evergreen tree native to Asia. The tree's ellipsoidal yellow fruit is used for culinary and non-culinary purposes throughout the world, primarily for its juice, though the pulp and rind (zest) are also used in cooking and baking. The juice of the lemon is about 5% to 6% citric acid, which gives lemons a sour taste. The distinctive sour taste of lemon juice makes it a key ingredient in drinks and foods such as lemonade.

Growing Lemons

Figure: *A lemon tree in Corsica*

Lemons are often grafted to more vigorous rootstocks.

Varieties

The Bonnie Brae is oblong, smooth, thin skinned and seedless; mostly grown in San Diego County. The Bush lemon tree, a naturalized lemon, grows wild in subtropical Australia. It is very hardy, and has a thick skin with a true lemon flavor; the zest is good for cooking. It grows to about 4 m in a sunny position.

The Eureka grows year-round and abundantly. This is the common supermarket lemon, also known as "Four seasons" (*Quatre Saisons*) because of its ability to produce fruit and flowers together throughout the year. This variety is also available as a plant to domestic customers.

The Femminello St. Teresa, or Sorrento is native to Italy. This fruit's zest is high in lemon oils. It is the variety traditionally used in the making of limoncello.

The Jhambiri (*C. jhambiri*), also known as rough lemon and bush lemon, has a rough skin, lemon yellow exterior and a very sour pulp. It is widely used as a rootstock in South Asia.

The Lisbon is a good quality bitter lemon with high juice and acid levels, the fruits of Lisbon are very similar to Eureka. The vigorous and productive trees are very thorny, particularly when young.

The Meyer lemon is a cross between a lemon and possibly an orange or a mandarin, and was named after Frank N. Meyer, who first discovered it in 1908. Thin-skinned and slightly less acidic than the Lisbon and Eureka lemons, Meyer lemons require more care when shipping and are not widely grown on a commercial basis. Meyer lemons have a much thinner rind, and often mature to a yellow-orange colour. Meyer lemons are slightly more frost-tolerant than other lemons.

The Ponderosa lemon is more cold-sensitive than true lemons; the fruit are thick-skinned and very large. It is likely a citron-lemon hybrid.

The Variegated Pink is a varietal of the eureka or lisbon cultivars with variegated patterns in the foliage and the rinds of immature green fruit. Upon maturing to yellow, the variegated pattern recedes in the fruit rind. The flesh and juice are pink or pinkish-orange instead of yellow.

The Verna is a Spanish variety of unknown origin.

The Yen Ben is an Australasian cultivar.

Culinary Uses

Figure: *Lemon Pickle*

Lemon juice, rind, and zest are used in a wide variety of food and drink. Lemon juice is used to make lemonade, soft drinks, and cocktails.

It is used in marinades for fish, where its acid neutralizes amines in fish by converting them into non-volatile ammonium salts, and meat, where the acid partially hydrolyzes tough collagen fibres, tenderizing the meat, but the low pH denatures the proteins, causing them to dry out when cooked. Lemon juice is frequently used in the United Kingdom to add to pancakes, especially on Shrove Tuesday.

Lemon juice is also used as a short-term preservative on certain foods that tend to oxidize and turn brown after being sliced, such as apples, bananas and avocados, where its acid denatures the enzymes that cause browning and degradation.

Lemon juice and rind are used to make marmalade and lemon liqueur. Lemon slices and lemon rind are used as a garnish for food and drinks. Lemon zest, the grated outer rind of the fruit, is used to add flavor to baked goods, puddings, rice and other dishes.

Preserved lemons are a part of Moroccan cuisine. They are also one of the main ingredients in many Indian cuisines. Either lemon pickle or mango pickle is part of everyday lunches in Southern India.

The leaves of the lemon tree are used to make a tea and for preparing cooked meats and seafoods.

Industrial

Lemons were the primary commercial source of citric acid before the development of fermentation-based processes.

As a Cleaning Agent

The juice of the lemon may be used for cleaning. A halved lemon dipped in salt or baking powder is used to brighten copper cookware. The acid dissolves the tarnish and the abrasives assist the cleaning. As a sanitary kitchen deodorizer the juice can deodorize, remove grease, bleach stains, and disinfect; when mixed with baking soda, it removes stains from plastic food storage containers. The oil of the lemon's peel also has various uses. It is used as a wood cleaner and polish, where its solvent property is employed to dissolve old wax, fingerprints, and grime. Lemon oil and orange oil are also used as a nontoxic insecticide treatment.

A halved lemon is used as a finger moistener for those counting large amounts of bills, such as tellers and cashiers.

Medicinal

Lemon oil may be used in aromatherapy. Researchers at The Ohio State University found that lemon oil aroma does not influence the human immune system, but may enhance mood. The low pH of juice makes it antibacterial, and in India, the lemon is used in Indian traditional medicines Siddha Medicine and Ayurveda.

Other

One educational science experiment involves attaching electrodes to a lemon and using it as a battery to produce electricity. Although very low power, several lemon batteries can power a small digital watch. These experiments also work with other fruits and vegetables. Lemon juice is also sometimes used as an acid in educational science experiments.

Lemon juice may be used as a simple invisible ink, developed by heat.

Pest Management

Monitoring Systems: Field monitoring is the key to an integrated pest management programme. When performed systematically on a quantitative basis, it provides the information needed to detect important changes in pest activity and disease incidence; evaluate pesticide performance; and establish seasonal records for evaluating and upgrading managerial decisions. When weather data and cultural practices that influence pest behaviour are considered, pest behaviour and populations can better be anticipated.

Monitoring systems may be limited to a single primary pest Such as the citrus rust mite, and to environmental conditions that influence rust mite behaviour or they may include all major pests and beneficials. Monitoring techniques vary depending on individual preferences and management intensity, but should provide a consistent, repeatable means to numerically measure pest activity. Follow monitoring procedures upon which the control action thresholds were designed.

Monitoring Frequency

Monitoring frequency depends on the pests' reproductive capacities and time needed under optimum conditions for populations to reach economic damage thresholds. Because of the reproductive potential of citrus rust mite, biweekly monitoring of susceptible fruit is required from petal fall until harvest or arrival of cool weather. Other pests generally require less frequent monitoring. Outside of the petal-fall to harvest interval, monthly monitoring is adequate for the citrus mite complex and other potential pest or disease situations.

Sampling Methods

Sample size, location and selection are important to any sample programme. Each citrus orchard is unique and requires individual monitoring; the first step in developing a monitoring system is to decide how many trees need to be examined and, within this parametre, how many fruit, leaf or soil samples need to be observed or collected for adequate representation and sampling reliability. Sampling may involve trees randomly selected for each monitoring period or permanently designated station trees. Trees should be dispersed sufficiently to achieve representative sampling of the entire orchard. This can be accomplished in most rectangular orchards by sampling in a diagonal direction. For square orchards, a Z-shaped pattern is appropriate.

The station tree concept involves selection of representative trees at the time of orchard establishment or at the beginning of each season. Selected trees are flagged, numbered and revisited each monitoring period throughout the season. Random samples provide information on a larger proportion of trees within an orchard over time. Station trees more closely follow pest or disease dynamics on a few individual trees. Both systems have merit and, depending on the desired result and amount of time to be invested, a combination of both methods may prove most useful.

Field Records

Systematic recording of data in the field is time consuming but necessary for clarity in review and interpretation of results. If field counts are considered impractical and some form of rating scale is employed (e.g., trace, low, medium, high, very high), the assignment of numerical values from 1-5 is acceptable. Organization of the data in summary form following field sampling allows for id review and assessment of individual orchard situations. Information gathered during the monitoring process aids in planning future pest management strategies.

Control Action Criteria

Control action criteria can be defined as specific pest population levels or threshold levels that signal impending economic damage to the crop or trees. Ideally, each threshold value considers the pest's life cycle, its potential for rapid population increases under existing or anticipated environmental conditions, orchard history, the type of damage likely to occur in proportion to control costs, other stresses the trees are experiencing and other relevant factors.

Progress has been made in establishing threshold values for citrus rust mite, the major economic pest of Valley citrus, but standardized criteria for other common pests are lacking. Suggestions in this text for key pests are based on guidelines developed in other citrus production states and observational data recently accumulated by local IPM programmes. Action criteria in any pest management programme are flexible and should be altered to fit individual management conditions.

Monitoring Guidelines

These guidelines provide information on how to monitor specific pests and offer insight on what constitutes reason for control action. Pre-season monitoring suggestions are provided to encourage early season inventory of pest populations and disease potential in advance of post-bloom control effects. Early monitoring provides information helpful in selecting the most appropriate and cost-effective pesticides. This early information may be useful in making volume purchases of pesticide materials and for determining which orchards merit immediate attention.

Citrus Mite Complex

Field studies of rust mite behaviour have indicated that populations in excess of 50 mites per square centimetre (cm^2) on fruit for 2 weeks or more are needed for rind damage to become visible. Also, rust mite populations on infested fruit may more than quadruple in a 2-week period under favourable conditions. These data support the need for a 2 week monitoring frequency for rust mite and are useful in establishing action thresholds.

To avoid rust mite management decisions being biased by two or three heavily infested fruits, consider the proportion of infested fruit. Also, consider mite distribution within trees or within orchards. Under this particular sampling regime, less than 20 percent incidence of infested fruit or of station trees would ameliorate average rust mite density values, and an incidence of more than 20 percent would substantiate or override them. The following criteria are utilized:

- Less than three mites /cm^2 and less than 20 percent of sample fruit with mites-no action necessary.
- Three to five mites/cm^2 and 20 percent of sample fruit with mites-borderline situation.
- More than five mites/cm^2 and more than 20 percent of sample fruit with mites-control should be initiated.

A sampling technique involving different action criteria has been used successfully by growers for several years. It involves random sampling of a single fruit per tree on 50 randomly selected trees within an orchard. Rust mite counts are expressed in numbers per cm^2. Action criteria vary according to irrigation schedules and weather outlook, but generally action is indicated when average mite density exceeds 20 mites/cm^2 and/or 25 percent of sample fruit are observed to be infested.

Factors to consider when making control decisions regarding rust mites include:

- Variety-given similar rust mite densities, oranges are downgraded less by mite damage than grapefruit.
- Lead time needed by producer to spray.
- Weather outlook and time of year-heaviest rainfall periods are May/June and August /September. Evaluate borderline mite densities during these periods in light of possible re-entry difficulties.
- Irrigation scheduling-orchards expected to need water within 10 to 14 days of an anticipated action threshold should be sprayed prior to irrigation.
- Miticide used in previous spray treatment-mite populations may rebound quickly following use of low residual materials.
- Orchard history-certain orchards, because of location, cultural practices, tree spacing or lack of uniformity in tree sizes and other reasons, consistently experience rapid re-infestation following control efforts.

Citrus red mites feed on tender new leaves and populations increase with new growth flushes. Population levels decline during hot, dry July/ August weather and resurge with the fall flush.

Control action guidelines for red mite on Texas citrus are incomplete pending further research on natural control agents and tree tolerance levels. Factors to consider in evaluating need for red mite control include:

- General tree vigor.
- Mite reproductive potential and environmental conditions.

Texas citrus mites do not require specific control action in most seasons if trees and fruit are protected from other mite and insect pests. Threshold levels vary according to tree vigor and should be determined by period during which heavy feeding is sustained rather than by specific mite numbers. Numbers exceeding 10 adult mites per leaf over a period of 30 days have been associated with winter defoliation of crown leaves.

Major considerations in Texas citrus mite management include:

- Tree vigor.
- Current and projected climatic conditions.

- Deleterious effect of fungicides applied for melanose or rust mite control on beneficial fungi.
- Effect of certain fall scalicide applications on over-wintering mite populations (azinphos-methyl and methidathion have been implicated with increased mite populations).
- Time of year-predisposition of mite-damaged leaves to defoliation by strong, dry winds greatest during fall/winter months.

False spider mites' increase in the past was associated with the use of organophosphate pesticides to control other pests and situations involving use of zineb alone for control of rust mite. With the present trend to use specific miticides such as dicofol, chlorobenzilate and fenbutatin-oxide, primarily for rust mite control, recent false spider mite problems have been associated with orchards where aldicarb soil treatment replaced post-bloom and late spring sprays. In aldicarb-treated orchards where the need for rust mite control may not arise until late summer, start monitoring for false spider mite in July and continue through September.

Loquat

The loquat (*Eriobotrya japonica*) is a species of flowering plant in the family Rosaceae, native to south-central China. It is a large evergreen shrub or small tree, grown commercially for its yellow fruit, and also cultivated as an ornamental plant.

Eriobotrya japonica was formerly thought to be closely related to the genus *Mespilus*, and is still sometimes known as the Japanese medlar. It is also known as Japanese plum and Chinese plum.

Eriobotrya japonica is a large evergreen shrub or small tree, with a rounded crown, short trunk and woolly new twigs. The tree can grow to 5–10 metres (16–33 ft) tall, but is often smaller, about 3–4 metres (10–13 ft). The leaves are alternate, simple, 10–25 centimetres (4–10 in) long, dark green, tough and leathery in texture, with a serrated margin, and densely velvety-hairy below with thick yellow-brown pubescence; the young leaves are also densely pubescent above, but this soon rubs off.

Fruit

Loquats are unusual among fruit trees in that the flowers appear in the autumn or early winter, and the fruits are ripe in late winter or early spring. The flowers are 2 cm (1 in) in diametre, white, with five petals, and produced in stiff panicles of three to ten flowers. The flowers have a sweet, heady aroma that can be smelled from a distance.

Loquat fruits, growing in clusters, are oval, rounded or pear-shaped, 3–5 centimetres (1–2 in) long, with a smooth or downy, yellow or orange, sometimes red-blushed skin. The succulent, tangy flesh is white, yellow or orange and sweet to subacid or acid, depending on the cultivar.

Figure: *Fruit structure*

Each fruit contains from one to ten ovules, with three to five being most common. A variable number of the ovules mature into large brown seeds. The skin, though thin, can be peeled off manually if the fruit is ripe. In Egypt varieties with sweeter fruits and fewer seeds are often grafted on inferior quality specimens.

The fruits are the sweetest when soft and orange. The flavour is a mix of peach, citrus and mild mango.

The loquat is originally from southeastern China. It was introduced into Japan and became naturalised there in very early times, and has been cultivated there for over 1,000 years. It has also become naturalised in Afghanistan, Australia, Bermuda, Chile, Kenya, India, Iraq, South Africa, the whole Mediterranean Basin, Pakistan, New Zealand, Réunion, Tonga, Central America, Mexico, South America and in warmer parts of the United States (Hawaii, California, Texas, Louisiana, southern Georgia and Florida). Chinese immigrants are presumed to have carried the loquat to Hawaii.

The loquat was often mentioned in ancient Chinese literature, such as the poems of Li Bai. In Portuguese literature, it is mentioned since before the Age of Discovery.

Cultivation

Over 800 loquat cultivars exist in Asia. Self-fertile variants include the 'Gold Nugget' and 'Mogi' cultivars. The loquat is easy to grow in subtropical to mild temperate climates where it is often primarily grown as an ornamental plant, especially for its sweet-scented flowers, and secondarily for its delicious fruit. The boldly textured foliage adds a tropical look to gardens, contrasting well with many other plants. It is popular in the American South as an ornamental plant for its blossoms, though winter frosts rarely allow the flowers to survive and bear fruit the following spring.

Figure: *Loquat in flower. This is a cultivar intended for home-growing, where the flowers open gradually resulting in fruit also ripening gradually.*

There are many named cultivars, with orange or white flesh. Some cultivars are intended for home-growing, where the flowers open gradually, and thus the fruit also ripens gradually, compared to the commercially grown species where the flowers open almost simultaneously, and the whole tree's fruit also ripens together.

Japan is the leading producer of loquats followed by Israel and then Brazil.

In temperate climates it is grown as an ornamental with winter protection, as the fruits seldom ripen to an edible state. In the United Kingdom, it has gained the Royal Horticultural Society's Award of Garden Merit.

In the highland parts of Central America, the loquat has become naturalized, and is often found growing wild in areas that have been disturbed but abandoned, its seeds having been dispersed by birds.

Below 1000 metres, the fruit remains inedible for its high acidity, but above it, the wild fruit is appreciated and much harvested for its sweet, fruity flavor. It is occasionally planted for living fenceposts, as the tree is long-lived, not much subject to disease, and the wood is hard and durable. Good quality logs are much sought-after by furniture makers in Central America, who prize its hardness and durability.

In the US, the loquat tree is hardy only in USDA zones 8 and above, and will flower only where winter temperatures do not fall below 30 °F ("1 °C). In such areas, the tree flowers in autumn and the fruit ripens in late winter.

Culinary Use

The loquat has a high sugar, acid, and pectin content. It is eaten as a fresh fruit and mixes well with other fruits in fresh fruit salads or fruit cups. The fruits are also commonly used to make jam, jelly, and chutney, and are often served poached in light syrup. Firm, slightly immature fruits are best for making pies or tarts.

The fruit is sometimes canned. The waste ratio, however, is 30 percent or more, due to the seed size.

The fruit is also processed into confectioneries.

Loquats are abundant in Pakistan, from Islamabad north, during the month of April, where the sour unripe fruit are used to make chutneys and sauces.

Alcoholic Beverages

Loquats can also be used to make light wine. It is fermented into a fruit wine, sometimes using just the crystal sugar and white liquor. Lemon or lemon zest is often paired with the wine because the fruit has very low acidity.

In Italy nespolino liquer is made from the seeds, reminiscent of nocino and amaretto, both prepared from nuts and apricot kernels. Both the loquat seeds and the apricot kernels contain cyanogenic glycosides, but the drinks are prepared from varieties that contain only small quantities (such as Mogi and Tanaka), so there is no risk of cyanide poisoning.

Nutrition

The loquat is low in saturated fat and sodium, and is high in vitamin A, dietary fibre, potassium, and manganese.

Like most related plants, the seeds (pips) and young leaves of the plant are slightly poisonous, containing small amounts of cyanogenic glycosides (including amygdalin) which release cyanide when digested, though the low concentration and bitter flavour normally prevents enough being eaten to cause harm.

Medicinal

Loquat syrup is used in Chinese medicine for soothing the throat and is a popular ingredient for cough drops. The leaves, combined with other ingredients and known as *pipa gao* (‡gwg ; pinyin: pípágâo; literally "loquat paste"), it acts as a demulcent and an expectorant, as well as to soothe the digestive and respiratory systems.

In Japan, loquat leaves are dried to make a mild beverage known as *biwa cha* by brewing them using the traditional Japanese *senjiru* method. *Biwa cha* is held to beautify skin and heal inflammatory skin conditions such as psoriasis and eczema and to heal chronic respiratory conditions such as bronchitis.

Eaten in quantity, loquats have a gentle but noticeable sedative effect, lasting up to 24 hours.

5

Arid and Semi-arid Fruits

Date Palm

Phoenix dactylifera (date, or date palm) is a palm in the genus *Phoenix*, cultivated for its edible sweet fruit. Although its place of origin is unknown because of long cultivation, it probably originated from lands around the Persian Gulf. The species is widely cultivated and is reportedly naturalized in Australia, Spain, North Africa, the Canary Islands, Madeira, Cape Verde, the Sahel region of Africa, Mauritius, Réunion, Afghanistan, Pakistan, India, China (Fujian, Guangdong, Guangxi, Yunnan), Fiji, New Caledonia, the United States (California, Nevada, Florida), Puerto Rico, northern Mexico, El Salvador, the Leeward Islands, the Cayman Islands, and the Dominican Republic.

Phoenix dactylifera grows 70–75 feet (21–23 m) in height, growing singly or forming a clump with several stems from a single root system. The leaves are 4–6 metres (13–20 ft) long, with spines on the petiole, and pinnate, with about 150 leaflets; the leaflets are 30 cm (12 in) long and 2 cm (0.79 in) wide. The full span of the crown ranges from 6–10 m (20–33 ft). Dates contain 20–70 calories each, depending on size and variety.

Dates have been a staple food of the Middle East and the Indus Valley for thousands of years. They are believed to have originated around Iraq, and have been cultivated since ancient times from Mesopotamia to prehistoric Egypt, possibly as early as 4000 BCE. The Ancient Egyptians used the fruits to make date wine, and ate them at harvest. There is archaeological evidence of date cultivation in eastern Arabia in 6000 BCE. (Alvarez-Mon 2006).

There is also archeological evidence of date cultivation in Mehrgarh around 7000 BCE, a Neolithic civilization in what is now western Pakistan. Evidence of cultivation is continually found throughout later civilizations in the Indus Valley, including the Harappan period 2600 to 1900 BCE.

In later times, traders spread dates around South West Asia, northern Africa, and Spain and Italy. Dates were introduced into Mexico and California by the Spaniards in 1765, around Mission San Ignacio.

A date palm cultivar, known as Judean date palm is renowned for its long-lived orthodox seed, which successfully sprouted after accidental storage for 2000 years. This particular seed is presently reputed to be the oldest viable seed but the upper survival time limit of properly stored seeds remains unknown.

Fossil records show that the date palm has existed for at least 50 million years.

Dates

The fruit is known as a date. The fruit's English name (through Old French), as well as the Latin species name *dactylifera*, both come from the Greek word for "finger", *dáktulos*, because of the fruit's elongated shape. Dates are oval-cylindrical, 3–7 cm long, and 2–3 cm (0.79–1.18 in) diametre, and when ripe, range from bright red to bright yellow in colour, depending on variety. Dates contain a single stone about 2–2.5 cm (0.79–0.98 in) long and 6–8 mm (0.24–0.31 in) thick. Three main cultivar groups of date exist: soft (e.g. 'Barhee', 'Halawy', 'Khadrawy', 'Medjool'), semi-dry (e.g. 'Dayri', 'Deglet Noor', 'Zahdi'), and dry (e.g. 'Thoory'). The type of fruit depends on the glucose, fructose and sucrose content.

The date palm is dioecious, having separate male and female plants. They can be easily grown from seed, but only 50 percent of seedlings will be female and hence fruit bearing, and dates from seedling plants are often smaller and of poorer quality. Most commercial plantations thus use cuttings of heavily cropping cultivars. Plants grown from cuttings will fruit 2–3 years earlier than seedling plants.

Dates are naturally wind pollinated but in both traditional oasis horticulture and in the modern commercial orchards they are entirely pollinated manually. Natural pollination occurs with about an equal number of male and female plants. However, with assistance, one male can pollinate up to 100 females. Since the males are of value only as pollinators, this allows the growers to use their resources for many more fruit producing female plants. Some growers do not even maintain any male plants as male flowers become available at local markets at pollination time. Manual pollination is done by skilled labourers on ladders. In some areas such as Iraq the pollinator climbs the tree using a special climbing tool that wraps around the tree trunk and the climber's back to keep him attached to the trunk while climbing. Less often the pollen may be blown onto the female flowers by a wind machine.

Parthenocarpic cultivars are available but the seedless fruit is smaller and of lower quality.

Dates ripen in four stages, which are known throughout the world by their Arabic names *kimri* (unripe), *khlal* (full-size, crunchy), *rutab* (ripe, soft), *tamr* (ripe, sun-dried).

Dates are an important traditional crop in Iraq, Arabia, and north Africa west to Morocco. Dates are also mentioned more than 50 times in the Bible and 20 times in the Qur'an. In Islamic culture, dates and yogurt or milk are traditionally the first foods consumed for Iftar after the sun has set during Ramadan. Dates (especially Medjool and Deglet Noor) are also cultivated in America in southern California, Arizona and southern Florida in the United States and in Sonora and Baja California in Mexico.

Date palms can take 4 to 8 years after planting before they will bear fruit, and produce viable yields for commercial harvest between 7 to 10 years. Mature date palms can produce 68 to 176 kilograms (150 to 300 lb) of dates per harvest season, although they do not all ripen at the same time so several harvests are required. In order to get fruit of marketable quality, the bunches of dates must be thinned and bagged or covered before ripening so that the remaining fruits grow larger and are protected from weather and pests such as birds.

A large number of date cultivars are grown: The Gaza Strip, especially Deir al-Balah ("Village of Dates"), is known for its exceptionally sweet red dates. There are more than 100 known cultivars in Iraq. It should be noted, however, that a cultivar can have several names depending on the locality.

Fruit Food Uses

Dry or soft dates are eaten out-of-hand, or may be pitted and stuffed with fillings such as almonds, walnuts, pecans, candied orange and lemon peel, tahini, marzipan or cream cheese. Pitted dates are also referred to as *stoned dates*. Partially dried pitted dates may be glazed with glucose syrup for use as a snack food. Dates can also be chopped and used in a range of sweet and savory dishes, from tajines (tagines) in Morocco to puddings, ka'ak (types of Arab cookies) and other dessert items. Date nut bread, a type of cake, is very popular in the United States, especially around holidays.

Dates are also processed into cubes, paste called "'ajwa", spread, date syrup or "honey" called "dibs" or "rub" in Libya, powder (date sugar), vinegar or alcohol. Vinegar made from dates is a traditional product of the Middle East. Recent innovations include chocolate-covered dates and products such as sparkling date juice, used in some Islamic countries as a non-alcoholic version of champagne, for special occasions and religious times such as Ramadan. When Muslims break fast in the evening meal of Ramadan, it is traditional to eat a date first.

Dates can also be dehydrated, ground and mixed with grain to form a nutritious stockfeed. Dried dates are fed to camels, horses and dogs in the Sahara. In northern Nigeria, dates and peppers added to the native beer are believed to make it less intoxicating.

In Southeast Spain (where a large date plantation exists including UNESCO protected Palmeral of Elche) dates (usually pitted with fried almond) are served wrapped in bacon and shallow fried.

Nutritional Value

Dates provide a wide range of essential nutrients, and are a very good source of dietary potassium. The sugar content of ripe dates is about 80%; the remainder consists of protein, fibre, and trace elements including boron, cobalt, copper, fluorine, magnesium, manganese, selenium, and zinc. The glycemic index for three different varieties of dates are 35.5 (khalas), 49.7 (barhi) and 30.5 (bo ma'an).

The caffeic acid glycoside 3-O-caffeoylshikimic acid (also known as dactylifric acid) and its isomers, are enzymic browning substrates found in dates.

A 2008 study suggests that consumption of dates by women in the final four weeks of pregnancy has positive effects regarding labour. Women who had six dates daily were compared to women who did not eat dates. The date-consuming group had greater mean cervical dilation, higher proportion of intact membranes, higher proportion of spontaneous labour, lower use of prostin/oxytocin, and shorter latent phase of labour.

Other Uses of the Fruits

In Pakistan, a viscous, thick syrup made from the ripe fruits is used as a coating for leather bags and pipes to prevent leaking.

Uses of Other Parts of the Plant

Where craft traditions still thrive, such as in Oman, the palm tree is the most versatile of all indigenous plants, and virtually every part of the tree is utilized to make functional items ranging from rope and baskets to beehives, fishing boats, and traditional dwellings.

Seeds

Date seeds are soaked and ground up for animal feed. Their oil is suitable for use in soap and cosmetics. Date palm seeds contain 0.56–5.4% lauric acid. They can also be processed chemically as a source of oxalic acid. The seeds are also burned to make charcoal for silversmiths, and can be strung in necklaces. Date seeds are also ground and used in the manner of coffee beans, or as an additive to coffee. Experimental studies have shown that feeding mice with the aqueous extract of date pits exhibit anti-genotoxic and reduce DNA damage induced by N-Nitroso-N-methylurea.

Fruit Clusters

Stripped fruit clusters are used as brooms.

Sap

***Figure:** Sweet sap tapped from date palm in West Bengal, India*

In large parts of Northern India the local species of wild date palm, *Phoenix sylvestris*, is tapped for palm wine, while in Pakistan and other regional countries in the region it is now mostly tapped for jaggery and palm syrup production. Wild date palms are also tapped in large parts of Africa for palm wine. The process of palm tapping involves the cutting of the unopened flower stalk and then fastening a bottle gourd, clay or plastic vessel on to it. The palm sap then collects in the vessel and is harvested in the early morning hours. If a few drops of lime juice are added to the palm sap, fermentation can be stopped and the sap can then be boiled to form palm syrup, palm sugar, jaggery and numerous other edible products derived from the syrup. In India and Pakistan, North Africa, Ghana, and Côte d'Ivoire, date palms are tapped for the sweet sap, which is converted into palm sugar (known as *jaggery* or *gur*), molasses or alcoholic beverages. In North Africa the sap obtained from tapping palm trees is known as lâgbî. If left for a sufficient period of time (typically hours, depending on the temperature) lâgbî easily becomes an alcoholic drink. Special skill is required when tapping the palm tree so that it does not die.

Leaves

Date palm leaves are used for Palm Sunday in the Christian religion. In North Africa, they are commonly used for making huts. Mature leaves are also made into mats, screens, baskets and fans. Processed leaves can be used for insulating board. Dried leaf petioles are a source of cellulose pulp, used for walking sticks, brooms, fishing floats and fuel. Leaf sheaths are prized for their scent, and fibre from them is also used for rope, coarse cloth, and large hats. The leaves are also used as a lulav in the Jewish holiday of Sukkot.

Young date leaves are cooked and eaten as a vegetable, as is the terminal bud or heart, though its removal kills the palm. The finely ground seeds are mixed with flour to make bread in times of scarcity. The flowers of the date palm are also edible. Traditionally the female flowers are the most available for sale and weigh 300–400 grams. The flower buds are used in salad or ground with dried fish to make a condiment for bread.

Wood

Date palm wood is used for posts and rafters for huts; it is lighter than coconut and not very durable. It is also used for construction such as bridges and aqueducts, and parts of dhows. Leftover wood is burnt for fuel.

Cultural Symbols

The date palm represents the provincial tree of Balochistan (Pakistan) (unofficial).

Diseases

Date Palms are susceptible to a disease called Bayoud disease, which is caused by the fungus *Fusarium oxysporum*. This disease, which kills many of the popular older cultivars like Deglet Noor, has led to a major decline in production where it is present, notably Morocco and western Algeria. However, new cultivars resistant to the disease are being developed.

In the 1920s, eleven healthy Madjool palms were transferred from Morocco to the United States where they were tended by members of the Chemehuevi tribe in a remote region of Nevada. Nine of these survived and in 1935, cultivars were transferred to the US Date Garden in Indio, California. Eventually this stock was reintroduced to Africa and led to the US production of dates in Yuma, Arizona and the Bard Valley in California.

Fungal Diseases of Date Palm

Bayoud disease: Origin, distribution and economic importance

The name bayoud comes from the Arabic word, "abiadh", meaning white which refers to the whitening of the fronds of diseased palms. This disease was first reported in 1870 in Zagora-Morocco. By 1940, it had already affected several date plantations and after one century, the disease has practically affected all Moroccan palm groves, as well as those of the western and central Algerian Sahara (Killian and Maire, 1930; Toutain, 1967).

Bayoud disease causes considerable damage that can sometimes take on spectacular proportions when the disease presents its violent epidemic aspect. Bayoud has destroyed in one century more than twelve million palms in Morocco and three million in Algeria. Bayoud destroyed the world's most renowned varieties that are susceptible to the disease and particularly those which produce high quality and quantity fruit (Medjool, Deglet Nour, BouFegouss). It also accelerated the phenomenon of desertifi cation. The result is an infl ux of farmers who have abandoned their land and moved to large urban centres.

The continued spread of bayoud highlights the problem threatening the important plantations of Deglet Nour and Ghars in Oued Rhir, Zibans in Algeria and even in Tunisia, which is presently free of the disease, but has 70 % to 80 % of the date palm areas under varieties susceptible it..

The disease continues to advance relentlessly to the east, despite prophylactic measures and regular attempts at eradication undertaken in Algeria. It is evident therefore, that Bayoud constitutes a plague to

Saharan agriculture and at the present expansion rate, it will certainly pose serious problems of human, social and economic nature to other date-producing areas of the world.

Disease symptoms: The bayoud disease attacks mature and young palms alike, as well as offshoots at their base (Saaidi, 1979).

External symptoms: The first symptom of the disease appears on a palm leaf of the middle crown. This leaf takes on a leaden hue (ash grey colour) and then withens, from bottom to top, in a very particular way: some pinnae or spines situated on one side of the frond wither progressively from the base upward to the apex. After one side has been affected, the whitening begins on the other side, progressing this time in the opposite direction from the top of the frond to the base.

A brown stain appears lengthwise on the dorsal side of the rachis and advances from the base to the tip of the frond, corresponding to the passage of the mycelium in the vascular bundles of the rachis. Afterwards, the frond exhibits a characteristic arch, resembling a wet feather and hangs down along the trunk. This whitening and dying process of the pinnae may take from a few days to several weeks.

The same succession of symptoms then begins to appear on adjacent leaves. The disease advances ineluctably and the palm dies when the terminal bud is affected. The palm can die at any time from several weeks to several months after the appearance of the first symptoms. The rapid evolution of the symptoms depends mainly on planting conditions and on variety.

Internal symptoms: A small number of disease infected roots, reddish in colour, are revealed when an affected palm is uprooted. The spots are large and numerous towards the base of the stipe. As they advance towards the upper parts of the palm, the coloured conducting fascicles separate and their complicated path inside the healthy tissues can be followed.

Palm fronds manifesting external symptoms exhibit a reddish brown colour when cut, showing highly coloured conducting fascicles. There is, therefore, a continuity of vascular symptoms that exist from the roots of the palm to the tips of the palm fronds. The observation of symptoms is necessary to recognise the bayoud, but to identify this disease with certainty, samples of affected fronds must be analysed by a specialised laboratory.

Pathogen: The causal organism responsible for bayoud is a microscopic fungus which belongs to the mycofl ora of the soil and is named *Fusarium oxysporum forma specialis albedinis* (Killian and Maire, 1930; Malencon, 1934 and 1936).

Biology and Epidemiology

Survival *Fusarium oxysporum f. sp. albedinis* is preserved in the form of chlamydospores in the dead tissues of infected palm, especially in the roots which have been killed by the disease and in the soil.

Spread of Bayoud in palm groves Contamination occurs regularly from palm to palm and more rapidly as the amount of irrigation increases. The appearance of the disease in locations far from the original infected area is caused primarily by the transport of infected offshoots or palm fragments harbouring the fungus.

Host plants Many plants are often grown as intercrops in palm groves, notably lucerne (*Medicago sa-tiva L.*; alfalfa), henna (*Lawsonia inermis L.)* and vegetables. (Bult et al., 1967; Djerbi et al., 1985 and Louvet et al., 1973). These plants can harbour the bayoud organism without manifesting any symptoms (symptomless carriers).

Control of Bayoud Disease

Chemical control: Soil treatment of this type of disease is destined, *a priori*, to fail and should therefore be avoided. Chemical control can, however, be feasible in the event of the discovery of primary sources of infection in a healthy area. In this case eradication techniques should be used: palms are uprooted and incinerated on the spot. The soil is then treated with methyl bromide or chloropicrin and the area closed off with replanting prohibited until further notice.

Cultural control: Since the factors that favour high yield in date palms (irrigation, fertilisation, etc.) are the same that favour the growth of the fungus, cultural techniques are not advised. However, a signifi cant reduction in the amount of irrigation can retard the advance of infection,i.e. stopping irrigation between the months of May and October, during the hot season in the northern hemisphere (Pereau-LeRoy, 1958).

Since the contamination occurs mainly by root contact, disease-free palms can be isolated by digging a trench of 2 m deep around them. Water should be provided by a trough bridging the rest of the grove to this isolated plot. Under these conditions these palms can be protected for more than 10 years (Djerbi, 1983).

Prophylactic measures: The essential task is to prevent the movement of contaminated plant material from an infected palm grove to a healthy one. This material, as has been previously mentioned, consists mainly of offshoots, palm fragments, manure and infected soil, and artifacts made from these materials. Legislation preventing the conveyance of contaminated vegetative material from one country to another, or from one region to another, has been passed by various countries such as Algeria, Egypt, Iraq, Libya, Mauritania, Saudi Arabia, Tunisia and USA.

Genetic control; The only productive means of controlling bayoud disease lies in continued research into resistant varieties. Many resistant cultivars have already been obtained in Morocco from three sources: selection of bayoud-resistant varieties from those already existing (local and introduced), selection of high-quality, resistant clones from the natural population of the date palm, and creation of resistant and high quality varieties through a hybridisation programme (Djerbi et al., 1986; Toutain, 1968).

In addition, the present success of date palm propagation by in vitro culture will make it possible to rehabilitate the Moroccan and Algerian palm groves that have been destroyed by bayoud. It will also be possible to reconstitute the palm groves presently threatened by Bayoud and create new date-growing areas with the help of high quality, resistant varieties.

In conclusion, bayoud disease is an epiphytic disease for which there is no known cure at present. Only preventive measures could protect healthy date plantations from this disease. Therefore, the following measures are imperative:

- Forbid the introduction of offshoots and all other plant material (palm fragments, artifacts made from date material, manure and infected soil) originating from bayoud infected countries or regions.
- Forbid the import of seeds and unprocessed products of symptomless carriers such as Alfalfa (Lucerne) and Henna from bayoud-infected countries or regions.
- Adopt legislation preventing the conveyance of the above plant material.
- Immediately report cases where symptoms similar to the ones caused by the bayoud appear.
- Information on bayoud and other major diseases and pests is necessary for the success of all above actions and must be available to all date growers.

Black Scorch Disease

Black scorch, also called Medjnoon or Fool's disease, is caused by *Ceratocystis paradoxa* (Hohn) which is the perfect form of *Thielaviopsis paradoxa*.

Black scorch has been observed on date palm in all date growing areas of the world. Symptoms are usually expressed in four distinct forms: black scorch on the leaves, inflorescence blight, heart or trunk rot and bud rot on palms of all ages. Infections are all characterised by partial to complete necrosis of the tissues. Typical lesions are dark brown to black, hard, carbonaceous, and, as a mass, give the petioles, fruit strands and fruit stalks a scorched, charcoal-like appearance.

Decay is most serious when it attacks the terminal bud and heart leading to the death of the palm. Some palms recover, probably by development of a lateral bud from the uninjured portions of meristematic tissue. These palms show a characteristic bend in the region of infection. This is why it is called Medjnoon. They set normal growth back by several years.

According to Djerbi (1983), black scorch has been observed on 17 date varieties. Thoory, Hayani, Amhat, Saidy and Halawy varieties are highly susceptible. The disease has also been observed on Zahdi, Menakher, Baklany, Gantar, Halooa, Fteemy, Sukkar Nabat, Horra, Besser Haloo, Nakleh-Zianeh and Koroch varieties (Klotz and Fawcett, 1932). Medjool and Barhee varieties are also susceptible to the disease (Zaid's own observations).

Good sanitation is the first step in the control of black scorch. The affected fronds, leaf bases and inflorescences should be pruned, collected and immediately burned. The pruning cuts and surrounding tissues should be protected by spraying with Bordeaux mixture, lime-sulphur solution, copper sulphate lime mixture, dichlone, thiram or any new copper-based fungicides. Under a severe attack, affected palms are to be removed and burnt.

Brown Leaf Spot

Brown leaf spot as with other common date palm diseases, has also been observed in North Africa and the Middle East (Rieuf, 1968). Dark lesions are clearly delimited on green leaves, and on dying leaves the margin of the lesion remains reddish/brown as the centre becomes pale. Lesions also occur on the rachis, pinnae and spines. Brown leaf spot is caused by *Mycosphaerella tassiana* (De Not) Johns. Because it is a minor disease, no treatment is recommended. However, annual pruning of old infected leaves and their immediate burning is advised.

Diplodia Disease

Diplodia disease, caused by *Diplodia phoenicum* (Sacc), has been recorded on 20 date varieties all around the world, although it appears to be most common to Deglet Nour. Symptoms are severe on offshoots and are characterised by death either while they are still attached to the mother palm or after they have been detached and planted out. The fungus may infect the outside leaves and fi nally kill younger leaves and the terminal bud;, or the central cluster may be infected and die before the older leaves. Yellowish-brown streaks extend along the leaf base.

On the leaves of older palms, the ventral mid-portion of the stalks is commonly affected, showing yellowish brown streaks, 15 cm to over one metre in length, extending along the leaf base and rachis. The upper part of the leaves however, may still appear green and unaffected.

Since the fungus usually enters the palm through wounds made during pruning or cutting when removing the offshoots, one precaution is to disinfect all tools and cut surfaces. Dipping or spraying the offshoots with various chemicals (benomyl, Bordeaux mixture, methylthiophanate, thiram and other copper-based fungicides), has been found effective against the disease.

Graphiola Leaf Spot

Graphiola leaf spot is caused by *Graphiola phoenicis* (Moug) Poit., which is a smut fungus. It develops sub-epidermal, in small spots on both sides of the pinnae leaves, on the rachis and on the leaf base. The numerous fruiting structures emerge as small-yellow/brown to black sori, 1 to 3 mm in diametre, with two layers. These sori are abundant on three year-old leaves, conspicuous on two year-old, but absent or infrequent on one year-old leaves. This is because of the 10 - 11 month incubation cycle for this pathogen. On a leaf, sori are abundant on apical pinnae, less abundant on the middle section becoming even less on the basal section.

The normal 6 - 8 year life of date palm fronds will be reduced to 3 years by *Graphiola* disease and heavily infected leaves die prematurely which consequently reduce yield of the palm.

Graphiola leaf spot disease is most common in Egypt (Delta region and Fayum) but absent in the less humid oases. In Saudi Arabia, it is abundant in Kattif, Demam and Jeddah, but absent in Iraq. Reports of this disease also originate from Algeria and USA. Around the world it is the most widely spread disease and occurs wherever the date palm is cultivated under humid conditions - mostly marginal date growing areas (Mediterranean coast) but also in the southern most humid regions of Mali, Mauritania, Niger and Senegal.

Control measures include leaf pruning coupled with treatment with Bordeaux mixture or any large spectrum fungicide (mancozeb, cupric hydroxide, cupric hydroxide + maneb, or copper oxychloride + maneb + zineb; 3 to 4 applications on a 15-day schedule after, sporulation, have been recommended). Genetic tolerance has been found in some varieties (Barhee, Adbad, Rahman, Gizaz, Iteema, Khastawy, Jouzi and Tadala).

Khamedj Disease

Khamedj or infl orescencKe rot is a serious disease affecting most date growing areas of the old world. It causes damage on inflorescences in neglected palm groves in hot and humid regions, or in areas with prolonged periods of heavy rain, 2 to 3 months before emergence of spathes. The disease can reappear each year on the same palm with the same intensity and it is estimated that, in serious cases, 30 - 40 kg of fruits are lost annually (Chabrolin, 1928).

During 1948 - 1949 and 1977 - 1978 severe outbreaks occurred in Iraq at Basrah, affecting male and female palms and destroying 80 % of the harvest (Al Hassan and Waleed, 1977). Serious damage was also recognised in Katif in the Kingdom of Saudi Arabia in 1983, with losses ranging from 50 to 70 %.

The disease is caused by *Mauginiella scattae* Cav., which is always found in a pure state in affected tissues. However, *Fusarium moniliforme* and *Thielaviopsis paradoxa* may rarely cause inflorescence rot.

The first visible symptom of the disease appears on the external surface of unopened spathes and is in the form of a brownish or rusty-coloured area. It is most apparent on the internal face of the spathe where the fungus has already begun to infect the infl orescence. When the infected spathes split, they reveal partial or complete destruction of the flowers and strands. Severely damaged spathes may remain closed and their internal contents may be completely infected. The inflorescences become dry and covered with powdery fructifi cations of the fungus.

Transmission of the disease from one palm to the next occurs through the contamination of male inflorescences during the pollination period. The infection of the young inflorescence occurs early and happens when the spathe is still hidden in the leaf bases. The fungus penetrates directly into the spathe and then reaches the inflorescences where the fungus sporulates abundantly.

The frequent appearance of the disease in neglected date plantations indicates that good sanitation and effi cient maintenance is the first step in the control of Khamedj disease. The collection and burning of all infected inflorescences and spathes should be followed by treating the diseased palms with the following fungicides after the harvest and one month before the emergence of spathes: a bordeaux mixture or a copper (1/3), sulphate-lime (2/3) mixture or a 3 % dichlone spray or a 4 % thirame spray at the rate of 8 litres per palm or with benonyl and tuzet at the rate of 125 g/hl (Al Hassan et al., 1977).

Some varieties are particularly susceptible to Khamedj disease: Medjool, Ghars, Khadrawy and Sayer. Others manifest a good capacity for resistance: Hallawi, Zahdi, Hamrain and Takermest (Laville, 1973).

Omphalia Root Rot

Omphalia root rot was recorded in California, USA and in Mauritania by Fawcett and Klotz (1932) and Bliss (1944), respectively. It is also called a decline disease because of its association with declining date palms. Four Mauritanian varieties (Ahmar, Marsij, Mrizigueg and Tinterguel) were found to be susceptible to this disease by Sachs (1967). Unlike other date varieties planted in California, Deglet Nour was found to have the lower infection rate.

Two species of Omphalia (*O. tralucida* Bliss and *O. pigmentata* Bliss) cause the disease and are widely spread in date plantations of Coachella Valley, CA-USA and in Kankossa (Mauritania) (Djerbi, 1983).

The premature death of fronds followed by retardation and cessation of growth are the main disease characteristics followed by necrosis and destruction of the roots. A completely non- productive stage is the result of the attack. The use of Brestan or Dexon at the rate of one spray every two weeks for eight weeks was recommended by Sachs (1967) as a chemical control measure.

Belâat Disease

Belâat disease was reported by several authors and from several North African countries (Algeria, Morocco, Tunisia, etc.) (Maire, 1935; Monciero, 1947; Calcat, 1959 and Toutain, 1967). The entire cluster of young fronds will whiten and die as a result of the attack, followed by the infection and death of the terminal bud. Accompanied by secondary organisms, the infection will progress downward in the trunk as a conical wet heart rot form, releasing an odour of acetic and butyric fermentation.

Belâat disease is caused by *Phytophtora sp.* similar to *P. palmivora* (Djerbi, 1983). Effi cient maintenance of date plantations is highly recommended to avoid attacks by this disease. Spraying with maneb or Bordeaux mixture at the rate of 8 litres/palm could control the disease at its early stages. Offshoots of affected palms usually remain healthy.

Fruit Rot

Fruit rot damage varies from one year to another depending on humidity and rain and also on the time of these factors from the Khalal stage until fruit maturation. Even though losses vary from one country to another and from one variety to another, they can be easily estimated to be between 10 % and 50 % of the harvest The most common fungi causing fruit spoilage are the calyx-end rot caused by *Aspergillus niger* and the side spot decay caused by *Alternaria* sp.

Lowering the humidity inside the bunch, by the use of wire rings, and/or by removing a few fruit strands from the centre of the bunch, will facilitate ventilation and drying of wet fruit. Protection from rain or dew is reached by using paper covers in the early Khalal stage to cover the fruit bunch. Fungus spoilage could also be limited by dusting the fruit bunches during the Khalal stage with 5 % ferbam, 5 % malathion, 50 % sulphur and an inert carrier (40 %) (Djerbi, 1983).

Phytoplasmic Diseases of Date Palm

Lethal Yellowing

Lethal yellowing destroyed about 300,000 coconut palms in Miami (Florida, USA) in less than fi ve years (McCoy, 1976). Previously, the disease killed more than 15,000 coconut palms in Florida, (USA).

The host list of palm species attacked by lethal yellowing is large and includes *Phoenix dactylifera* L.; *P. canariensis* Hort., and *P. reclinata* Jacq. (Thomas, 1974).

Developing fruits of the coconut start dropping from the palm followed by the formation of new inflorescences which rapidly become necrotic. These first symptoms are followed by a rapid and generalised yellowing, leading to the death of the palm.

In date palm the fronds become desiccated and grey-brown instead of becoming yellow. A soft rot of the growing point occurs, converting the meristematic area into a putrid, slimy mass. The crown topples from the palm, leaving a naked trunk. The causal agent is a mycoplasma-like organism. It is believed that the pathogen is disseminated by wind-born arthropod vectors. Removal of diseased palms and their offshoots, quarantine measures, the use of tolerant types of palms and the treatment with antibiotics are the main control measures.

Al Wijam

Nixon (1954) observed this disease in Al Hassa (Saudi Arabia). In Arabic, Al Wijam means poor or unfruitful. The disease is characterised by a retardation in terminal bud growth,and the whole crown of leaves formed after the occurrence of the disease have the rosetting symptoms. Newly formed leaves are reduced in size and marked by a faint narrow, yellow longitudinal line on the midribs. Leaves become chloritic and their life span is reduced.

Death of leaves starts from the distal end and extends towards the base. Diseased spathes split open before their complete emergence and are reduced in size. The number and size of the bunches produced are also reduced year after year till the diseased palm fails to produce and dies.

Positive amplifi cation bands were obtained from DNA templates extracted from diseased tissue of date palm using the Polymerase Chain Reaction (PCR). These DNA tests offer basic support to the hypothesis that the cause of Al-Wijam disease is a Mycoplasma- like organism (Djerbi, 1999; personal communication).

Brittle Leaves Disease

Brittle leaves disease, also called "Maladie des Feuilles Cassantes" in French, was first observed in Nefta, Tozeur and Degache date plantations (Tunisia) and in Adrar, M'zab and Biskra (Algeria) (Djerbi, 1983).

Both adult and young palms including offshoots are attacked alike. A broad chlorotic striping of the pinnae followed by drying of the tip of the frond is the first symptom of this disease.

Yields drop signifi cantly as more fronds are affected and the retardation in terminal bud's growth becomes evident. Leaves are shorter and of irregular size.

The causal agent remains unknown and no fungi or other pathogens were isolated. However, recent investigations with PCR showed that the causal agent seems to be a Mycoplasma-like Organism (Djerbi, 1999; personal communication).

Chemical analysis of date palm leaves and soils showed that concentrations of all nutrients in the tissue were higher in leaves of unhealthy palms. The exception was the concentration of manganese, which was ten times lower in the unhealthy palms (Djerbi, 1983). In addition, the conductivity and the phosphorus concentrations of the soil with diseased palms are higher than that of healthy ones. These results suggest that the areas affected by the disease have a build-up of major nutrients and salts as a result of irrigation, which have contributed to the high electrical conductivity. High pH and conductivity may have caused lack of manganese in the soil.

Quarantine measures seem to be the only means of limiting the spread of the disease. Since manganese is defi cient in unhealthy palms, this nutrient could be brought to these palms either by spraying or by injection. Djerbi (1983) found a gradient of susceptibility within Tunisian varieties even though they all seemed to be equally attacked.

Diseases of Unknown Cause of Date Palm

Bending Head

Also called "Le Coeur qui penche" in French, the bending head is a minor disease observed in Algeria, Egypt, Mauritania and Tunisia (Munier, 1955). The central cluster of fronds takes the form of an erect fascicle with a bent tip. The trunk bends and may even break.

Thielaviopsis paradoxa and *Botryodiplodia theobromae* Pat are fungi commonly isolated from declining palms (Brun and Laville, 1965). Effi cient maintenance and appropriate sanitation of the date plantation is the first control measure. Diseased parts of infested palms are to be collected and burnt in order to limit the spread of the disease.

Dry Bone

Originally this disease was first reported by Fawcett and Klotz (1932) in USA. Other cases were found in Algeria, Egypt and Tunisia (Djerbi, 1983). According to Djerbi, the disease is characterised by whitish, irregular blotches and streaks on the leaf stalks, midribs and pinnae that become outlined by reddish brown margins. The name "dry bone" comes from the drying out of the surface of the leaf stalk with a hard, smooth and white appearance. Lesions, from one to several centimetres, involve only the epidermis and a thin layer of subjacent tissue.

According to Fawcett and Klotz (1932), a bacterium is commonly found associated with the lesions, and certain palms are more susceptible than others.

Faroun Disease

Laville and Sachs (1967) reported this disease of unknown cause from Mauritania. Affected palms, present a parasol form produced by the old and mid-level fronds, while new fronds present a short rachis with an irregular arrangement of pinnae and spines. Leaves remain green during the first stages and then decline and become yellow. The terminal bud assumes a conical form and becomes a stunted rosette.

All these symptoms are accompanied by the abortion of the axillary buds, resulting in failure of fl owering for one or two seasons before foliage symptoms appear. Two to four years is the average duration of the disease from the appearance of the symptoms to the death of the palm. According to Djerbi (1983) no varietal resistance has been observed.

Rhizosis

Also called "Rapid decline", rhizosis is a minor but fatal disease of unknown cause. The first symptom is premature falling off of fruits. However, if the attack is sometime after fruit development, the fruit withers and shrivels on the bunch. A reddish-brown discolouration of pinnae appears on mature fronds and the disease progresses from the bottom to the top of the fronds which rapidly die.

Offshoots die with the diseased mother palm and the disease is hence self-limiting. According to Djerbi (1983), no varietal resistance has been observed.

Physiological Disorders of Date Palm

Blacknose

Blacknose applies to the abnormally shrivelled and darkened tip of a date. Deglet Nour and Hayani seem to be the most susceptible varieties to this physiological disorder (Fawcett and Klotz, 1932).

Blacknose results from excessive checking of the epidermis, especially in the form of numerous small, transverse checks or breaks at the stylar end of the fruit. Pronounced shrivelling and darkening occur in proportion to the abundance of the checks and are related to humid weather at the Khalaal stage.

Given the fact that checking is induced by high humidity and rainfall, it follows that measures to avoid conditions that tend to increase humidity are to be taken. The conditions to be avoided include excessive soil moisture and the presence of intercrops and weeds, especially at the susceptible stage of fruit development. According to Nixon (1932), bagging the fruits in brown wrapping paper was found to inhibit the occurrence of blacknose checking. Over thinning can also increase the incidence of checking and subsequent development of blacknose.

Crosscuts

Crosscuts is a physiological disorder of fruit stalks and fronds reported from the United States, Pakistan and a few Middle East date growing countries such as Israel and Iraq (Bliss, 1937; Djerbi, 1980). In the United States more than 1,000 fruit bunches were damaged in a single plantation in 1934, up to a quarter of the crop was lost.

Crosscuts, or V- cuts, are clean breaks in the tissues of the fruit stalk bases and on fronds. It consists of a slight to deep notch, similar to a cut artifi cially done by a knife. Fruits borne on strands in line with the break wither and fail to mature properly. Crosscuts result from an anatomical defect in the fruit stalks and fronds involving internal, sterile cavities leading to mechanical breaks during elongation of the stalk or the fronds. Crosscuts are commonly found in varieties having crowded leaf bases and its incidence increases as the palms get older. Sayer and Khadrawy varieties are especially susceptible to this disorder, and are no longer propagated in some countries (Carpenter, 1975).

Crop losses may be avoided by using non-susceptible varieties, or by reducing the number of fruit stalks in susceptible varieties.

Whitenose

Whitenose disease is commonly found in Iraq, Libya and Morocco (Hussain, 1974; Djerbi, 1983). Dry and prolonged wind in the early Rutab stage causes rapid maturation and desiccation of the fruit resulting in whitish drying at the calyx end of the fruit. The affected fruit becomes very dry, hard and has a high sugar content. Hydration may correct this condition in harvested fruits.

Barhee Disorder

Barhee disorder is characterised by an unusual bending of the crown of Barhee variety. The disease was first reported in California (USA) by Darley et al. (1960) and later in Al Basra (Iraq) by Hussain (1974). It was also found at the Kibbutz YOTVATA (Israel) by Zaid (1996). Affected palms were found to bend mostly to the south and sometimes to the south-west.

At the Kibbutz Kineret (Israel), this phenomenon is severe and bending could reach an angle of about 90°. In Israel this bending disorder is also found with Dayri variety. Literature shows that it also affects Jahla and Aguellid varieties (Djerbi, 1983).

Neither the cause nor the control of this disorder is known. However, at Yotvata Kibbutz (Israel), growers are correcting this situation by fi xing a heavy iron bar to the opposite side of the bending; fruit bunches from the opposite side are tied to this bar in order to move the actual weight against the bending side. It seems that within 2 to 3 years, the bending is corrected. Bunch handling is also proposed to correct such an abnormality (Yost, 1968).

Black Scald

Black scald, different from blacknose, is a minor disorder of unknown cause occurring in the United States (Djerbi, 1983). It consists of a blackened and sunken area with a defi nite line of demarcation. The disease usually appears on the tip or the sides of the fruit, and affected tissues have a bitter taste. The appearance of the disorder suggests exposure to high temperature, but the exact cause is not defi nitely known (Nixon, 1951).

Bastard Offshoot

This is a deformed growth of date palm vegetative buds especially of offshoots fronds. Mohamed and Al-Haidari (1965) stated that the bastard condition is due to infestation by the date palm bud mite *Makiella phoenicis* K. It may also be due to reduction in growth caused by an inequilibrium of growth regulators.

Leaf Apical Drying

This is not a disease but a physiological reaction to transplantation of adult palms (injury of their root system). All palms with these symptoms recover within two to three years after their transplanting.

Fertilisation Injury

This type of injury is present only with young tissue culture-derived palm plants (first two years after fi eld planting) and when fertilisers (N, P, K) are applied too close to the palm's stipe. The nature of fertilisers is not the cause, but rather how close to the stipe the fertiliser was applied. If the damage is severe, it could cause the death of the young palm.

Frost Damage

As stated in Chapter IV the date palm resists large temperature variations (-5 to 50°C) with a growth optimum between 32 and 38°C and a zero of vegetation of about 7°C. The vegetative activity will also decrease above 40°C and ceases around 45°C.

When temperature falls below 0°C, it causes serious metabolic disorders with some injury to date palm leaves characterised by a partial or total desiccation. Water of protoplasma freezes after coming out from the cells. During defrost, water invaded inter-cellular spaces and affected leaves turn brown and desiccated. The severity of damage is related to the intensity and duration of frost:

- At -6°C, leafl et ends become yellowish and dry up;
- At -12°C, leaves of external crown desiccate; and
- From -15°C, leaves of middle crown freeze and if low temperatures are suffi ciently prolonged, the central crown is reached and all foliage desiccates and the palm seems to be completely burned.

The relative stable temperature of terminal bud and trunk allows the date palm to resist frost in winter, and high temperature in summer. In fact, the terminal bud is protected by the fi brillium and the leaf bases; the internal temperatures of the trunk and terminal bud undergo less big variations than those of atmosphere; the difference is round 14°C less in summer and 12°C more in winter.

Frost injury to the date palm groves is not in direct loss of fruit on the palm but in freezing and loss of leaves so that the palm cannot support and mature the fruit crop the following year. Serious damage caused by frost was observed in plantations in Morocco and in USA where temperatures of approximately -15°C occurred and frost caused a complete desiccation of leaves. In Morocco, palms were considered lost and the damage looked like a disaster to the local population. However, in spring, terminal buds started to grow although they were severely affected, and a good bloom was obtained.

The most practical and available protection for the date growers is to turn on the water and keep the date plantation wet when the temperature begins to get low enough (-5°C and below). A date plantation just irrigated or being irrigated when the temperature falls, has some heat stored, which gives protection.

Data are also available on principal date varieties and their susceptibility to cold:

Resistant: Zahdi.

Moderatly susceptible: Bentamoda, Bentkbala, Besser Halou, Hayani, Itima, Jouze, Khastawi, Mesh Degla, Sayer, Tadala, Tazizot and Thoury.

Susceptible: Ammari, Amri, Arechti, Barhee, Beid Hmam, Dayri, Deglet Nour, Horra, Khadrawy, Maktoum, Medjool, Menakher and Saidy.

Highly susceptible: Brain, Fursi, Hallawy, Hilali, Khlass, Khush Zebda and Ghars.

Lack or Excess of Water

The growth of the date palm is highly affected by variations in water availability and the water content of the soil. A decrease in yield, or complete failure in fruit production could result from these water variations.

To compensate for high evapotranspiration, the date palm requires a quantity of water from 1,500 to 2,800 mm/year. Prolonged water stress will signifi cantly decrease growth and yield, and if the drought continues for several years, date palm can dry up and die.

On the other hand, when the water table is high and drainage is inadequate and/or the leaching and transport of soluble salts is not complete, high evaporation rates tend to increase the concentration of salts in soil and in surface water. However, there are limits of salt tolerance and the date palm will not grow when soluble salt of the soil is above 6 percent. As stated in Chapter IV, the following shows the relationship between salts, growth and yield:

- irrigation with water of salinity up to 3.5 mmhos/cm (i.e. 2240 ppm) will not affect the yield, provided that the leaching requirement of 7 % is provided for.
- With an irrigation water of 5.3 mmhos/cm salt content and a leaching requirement of 11 %, yield reduction is only 10 %.
- When the salt content of the irrigation water reaches 10 mmhos (i.e. 6400 ppm) and a leaching requirement of 21 %, the reduction in yield is around 50 %.

The timing of leaching must be adjusted in each case, according to the quantities of soil and water, conditions of drainage, and characteristics of rainfall.

Although date palms are resistant to fl ooding, healthy growth of palms requires a well-drained soil, and it is clear that irrigation must always go hand in hand with drainage.

Serious losses, sometimes irreversible may occur in neglected date plantations. In such cases signs of decline appear on palms favoured by root penetration of numerous saprophytes and parasites that could lead to the death of palms (Djerbi, 1983).

Major Pests of Date Palm

The date palm and its fruits are subject to attacks by several pests that are, in most cases, well adapted to the oasis environment. Damage caused by pests is considerable and leads to heavy economic losses.

White Scale

White scale, caused by *Parlatoria blanchardii* Targ., is widely present in most date palm growing areas of the world except in USA, where it was eradicated in 1936, and in some countries of the southern hemisphere (Namibia and RSA). It is considered a serious pest in Algeria, Kuwait, Libya, Mauritania, Morocco and Tunisia. Iraq, Oman, Saudi Arabia and Sudan consider this pest a moderate one, while Egypt, Jordan, UAE and Yemen consider it a minor pest.

Damage by white scale is very serious on young palms between two to eight years of age, but even under severe attacks, the palm and its offshoots do not die. Nymphs and adults suck the sap from the leafl et, midribs and the dates. Under each scale insect, a discoloured area appears on the leafl et. Heavy infestation causes leafl ets to turn yellow and contributes to the premature death of the fronds .

Respiration and photosynthesis are almost stopped resulting in early death of the infested leaf. Damage on fruits is easily noticeable and the production is not marketable. The cycle of *Parlatoria blanchardii* Targ. is summarised. The number of generations developed during one year varies from three to four depending on temperature.

The natural enemies of *Parlatoria blanchardii* are: *Hemisarcoptes malus, Chrysoperla vulgaris, Cardiastethus nazarenus, Coccinellidae* (29 species), *Nitidulidae* (5 species), *Mycetaeidae* (1 species), *Aphytis mytilaspidis, Cybocephalus nigriceps, Cybocephalus rufi frones, Chilocorus bipustulatus var. iraniensis* and *Chilocorus sp.*

Natural enemies and pruning normally keep pest populations at tolerable levels. In the 1970s the coccinellid *Chilocorus bibustulatus* var. *iraniensis* was introduced into Mauritania and Morocco, but permanent establishment failed and efforts were discontinued. In the 1980s, attempts were made to introduce the coccinellids into northern Sudan, but they were not successful either. In 1993 the coccinellids were released in Oman, but there is no information on their establishment. The introduction of coccinellids is currently being investigated in Tunisia.

Chemical control appears to be conducted occasionally in young plantations. Mineral oils are used (Djerbi, 1994).

Red Scale

Red scale, *Phoenicococcus marlatti*. cockerell, is exclusively a pest of palms, particularly date palms, with other palms as host plants (e.g.: Doupalm, Canary Island palm and the California fan palm). It is probably found wherever date palm is cultivated, but with no great threat (Dowson, 1982). The extent of its damage is known to be less than that caused by the *Parlatoria* scale.

Leaves of date palm are often found to be clotted over with thin, minute, greyish scales with darker centres. The darker spot is oval in outline and is the body of the insect itself. The individual scale is seldom larger than a small pinhead, roundish in shape, and deep pink to dark red in colour, but partly or entirely covered with a white waxy secretion that forms a cottony mass (Nixon and Carpenter, 1978).

All exposed portions of the palm can be attacked by the pest. Heavy infestations could cause complete coverage of the leaf surfaces by scales, which will result in interference with the metabolic functions of the plant. Attacked leaves and underlying tissues may be damaged to a depth of a few millimetres and will consequently be killed in severe cases.

The red date scale usually stays out of the light and is found massed on the white tissues at the bases of the leaves and fruitstalks, where it is protected by fibre and other leaf bases. Frequently, the scale is found on roots underground. The red scale is not as easily detectable as most other scales because of its natural tendency to hide. Red scale is not suspected until the base of the green leaf is cut and subsequently observed. Stickney *et. al.* (1950) provided a comprehensive study of the insect's biology.

P. marlatti. passes its lifecycle in a protective covering of wax that it secrets. The female produces numerous eggs under the protective scale. After the eggs hatch, the nymphs crawl out and move about freely, feeding at various positions. Once a suitable location on the host plant is selected, nymph's will insert their needle-like mouth parts to suck the sap. When they start to feed, layers of wax, forming the covering of the scale over the body, are secreted.

Soon after beginning to feed, adults will moult. Later on, males are incapable of feeding and will mate with the females and die. The female, once fertilised, increases rapidly in size and produces eggs before dying within the scale.

The pest breeds actively during the summer months and hibernation starts in early winter. A complete life-cycle takes approximately 55 days during summer and 158 days during winter. Three to fi ve generations could be found annually.

It is worth mentioning that the scale appears to cause considerable damage to plants growing under favourable conditions. Areas where the climate is milder or more humid may also face severe scale attacks.

Even though this scale insect is regarded insignifi cant, and with no economic impact, the first measure is to cut away all attacked leaves and burn them in order to stop the spread of the pest. Infested palms, offshoots or even tissue culture-derived plants, which are still at the hardening phase, must be sprayed with malathion 370 - 450 g or with parathion 120 g a.m. dissolved in 450 litres of water.

Since the scale is a sucking insect, the use of ultracide or dimenthoate when the pest is mobile is also recommended (Djerbi, 1994). Infested offshoots could also be subjected to a temperature of 50°C for 65 hours in an insulated room. General predators, such as *Pharoscymnus anchorago* (Fairmaire), are considered as active predators.

Bou Faroua

Bou Faroua, also called Goubar or Old World date mite, is caused by *Oligonychus afrasiaticus* McGregor, and *O. pratensis* Banks. This mite is present in all date growing areas, and damage is severe in neglected plantations. Immediately after fruit set (Hababouk stage), mite eggs are deposited to produce larvae which will feed on the fruits and later cover these with a web retaining sand particles. The cycle length is about ten to fi fteen days depending on temperature. Mites will rapidly multiply causing the drop-off of the fruits. Affected mature fruits are of no commercial value.

Chemical analysis of infested and fully matured dates shows that the water soluble substances such as sugar are less in infested dates (Hussain, 1974). Under Iraq's climate, the Old World date mite has six overlapping generations during the fruiting season of palms (Hussain, 1974). The mite population on dates reaches its peak during the middle of July. The first appearance of mite on immature dates is during the first week of July. Even though they are found on all parts of the date, the majority of mites congregate near the calyx area, where most of the eggs are laid. Mite and eggs are also found on fruit stalks. The mites migrate to the palm crown during the last week of August. Hussain (1974) states that the fi bres and frond bases taken from infested palms during the winter months show adult and nymph mites. This mite does not hibernate on the leafl ets, date palm seedlings, offshoots or on the many species of vegetation in the plantation.

Dusting date bunches early in July with sulphur at the rate of about 100 - 150 g per palm is effective (Djerbi, 1994). The Iraqi variety "Sayer" is relatively resistant to mite attack.

Caroub Moth

Caroub moth, also called "Ver de la Datte" in French, is caused by *Ectomyelois ceratoniae*. Zeller, and is found in all date growing areas. The larva of the Caroub moth attacks dates in plantations, packing houses and stores. Eggs are laid on the dates and hatching begins four days later. The larval period is about three weeks in warm months and eight weeks in colder months. The pupal period is about fi ve days.

Taking into account the moth's life cycle, it is recommended to protect the fruit bunches, to clean the plantation from wind-fallen fruits and to fumigate harvested and stored dates. The use of pheromone traps will not only help to determine the emergence of moths but also to estimate the population level. The rate of infestation could be lowered by spraying the infested fruits with *Bacillus thuringiensis* (Djerbi, 1994).

Rhinoceros Beetle (Oryctes Rhinoceros Linné)

The adult beetle is a stoutly-built insect about fi ve centimetres in body length and shiny black in colour with a reddish under-surface covered with short, fi ne hair. Its tibiae are furnished with thorn-like spines. This insect has earned the name of rhinoceros beetle because of the presence on its head of a horn-like structure, which is conspicuously longer in the male.

The adults feed on tender leaves, inflorescences and fruit stalk of the fruit bunches of date palm, whereas the grubs thrive on decomposing dung and decaying vegetable matter like stumps and trunks of palms. This insect is also a pest of coconut and other palms.

Within a week of the emergence of the females they start laying eggs. The whitish-brown eggs are laid singly in dung heaps and decomposing vegetable matter. The eggs hatch out into fat soft-bodied pale-yellowish curled larvae in about 10 to12 days. The larvae become full-grown in about 4 or 5 months and they take another 6 to 7 months in hibernation before they transform themselves into pupae. The full-grown larva is a stout fl eshy creature measuring about 7 cm in length with brownish head and dirty white appearance. The full-fed grub pupates in the dung heaps, etc., in a specially prepared oval chamber made of soil or excretory matter. The adult beetles emerge from the pupae in about 3 to 4 weeks and fl y to nearby palms and start feeding on them causing damage. There is only one life-cycle during the year.

Contrary to other pests, only the adult beetles are responsible for causing damage to the palms. The pest has been found to be more destructive to young plants. They remain hidden during the daytime and become active at night, when they fl y about and reach the tops of date palms. They drill large holes close to the base of the growing heart-leaf and enter the stem. They feed on the softer tissues of the growing heart-leaf and cut right through it, with the result that further growth stops and the palm ultimately dies. The beetle also causes damage by boring into tender fronds, chewing tissues and throwing them out as a fi brous dry mass. Fronds may hence break and if the growing point is bored the plant dies off. Most of the damage occurs during the rainy season.

The adult beetles should be attracted and destroyed by putting up mercury-vapour light traps at regular intervals in infested plantations.

The light trap is based on the fact that some insects are very active at night and are attracted by the light. This method of mechanical control is presently included in Integrated Pest Management.

The degree to which insects are attracted varies according to the type of traps as well as to the nature and power of light. It was shown that the mercury-vapour light is the best tool to attract insects.

The advantages of using light traps are::

- to obtain information on the number of captured species;
- to predict the occurrence of an outbreak of an insect-pest; and
- to use it as a mechanical control method since it can reduce the number of insects as well as production losses.

The insect collector (D) should be half filled with diesel, kerosene or paraffin;.

Red Palm Weevil and African Palm Weevil

The red palm weevil (RPW), *Rhynchophorus ferrugineus* Oliv., also called the Indian palm weevil, is well known in the Middle East where it causes severe damage on date palms. The RPW was first noted in the Arabian Peninsula in the mid 1980's and in Egypt in 1992. The weevil was first observed in Rass El Khaima, United Arab Emirates in 1985. Approximately, 5 to 6 % of palms in the Middle East region are infested with the RPW with an annual rate of infection of about 1.9.

The African palm weevil (APW), (*Rhynchophorus phoenicis F.*), was found by Zaid at two date plantations, one in the RSA and one in Zimbabwe. To the author's knowledge, it is the first time that this pest has been reported to attack date palms (*Phoenix dactylifera L*). It is also the first time that the genus *Rhynchophorus* has been reported to attack date palms in RSA and Zimbabwe.

The APW is suspected to originate from a local palm host commonly called Lala Palm (*Hyphaene coriacea*). However, in general, this species is known to occur naturally in southern Africa and is also widely distributed in Africa. It attacks a variety of palms in the genera of *Phoenix*, *Elaeis*, *Borassus*, *Hyphaene* and *Raphia*. The biology of the APW (*R. phoenicis*) is well known and summed up in Lepesme's "Les Insectes des Palmiers".

Infestation is often not apparent until extensive damage has already been caused and the palms are beyond recovery. In these infested plantations, we were looking for wilted/yellow inner leaves. When the observer got closer, a characteristic rotting odour could smelt. Small round holes at the sites of removed offshoots were also a clear indication of the presence of the weevil. Chewed up date palm fi bres were extruded, and a brown fl uid was oozing out of the holes on the stem. Cocoon, weevil and pupal fi bres are frequently found in the palm leaf base.

The following control measures are highly recommended: quarantine, plantation sanitation, chemical treatment, regular surveys, pheromone mass trapping and the use of nematodes. Furthermore, the control of the red and African palm weevils requires all these steps which are of equal importance. Not respecting even one of these measures will lead to infestation of date plantations.

Quarantine

It is imperative that all imports of date palm offshoots from infested areas (Middle East and Asia) to uninfested areas be prohibited. Other imports of palms into uninfested areas are to be carefully screened and put in quarantine so as not to introduce another species of *Rhynchophorus* or even another strain of *R. phoenicis* into the region. Even within the sub region of a sub continent the movement of palm plant material must be monitored through effective quarantine regulations.

Plantation Sanitation

Prevention of the infestation is essential, and the practice of good cultural techniques will protect the date plantation from infestation by weevils. Date palms are not to be stressed and appropriate irrigation and fertilisation programmes are to be respected. Removal of offshoots is to be properly implemented and the cut surface on the mother palm treated with PVC paint or a copper sulphate product. Soil is to be put around the base of the palm to protect the cut.

Over 80 % of weevil infestation occurs at the base near the offshoots or where offshoots have been removed. Palms that are stressed or damaged are vulnerable to attack and semi-chemicals emanating from these palms attract adult weevils.

Sanitation measures, such as the removal of dead palms or palms beyond recovery, are essential, as they are the ideal breeding places for the rhinoceros beetles that generally pave the way for entry of the palm weevil into young palms. Wounding of the palms, like cutting steps into the stem to facilitate climbing should be avoided. When the leaves are pruned, the grubs may tunnel their way into the stem through the cut end of the periole where eggs will be laid. Treatment of cut surfaces with PVC paint will ensure the control of infestation. Heavily infested date palms that can not be saved and the first infested palms of a healthy plantation are to be uprooted, burnt and buried outside the plantation to a depth of one metre. Growers must make sure that all weevils in the destroyed palm are killed. Many people do not like to be aggressive with phytosanitation, because of the investment in the palms, but the cost - if a weevil epizootic gets going - can accumulate to the loss of the whole plantation. Cut stumps and useless parts of the palm need to be destroyed in order to kill the early stages of the weevil. The holes and cuts made by the rhinoceros beetle constitute a favourable entry point to the weevil. These rhinoceros beetles must be attracted and destroyed by putting up mercury vapour light traps at regular intervals in the plantation.

Chemical Treatment

In case the whole plantation is infested, the grower could extend the life of the palm and resulting production by practising the following:

- cuts and holes made by the rhinoceros beetle should be treated (potassium cyanide, carbon bisulphate, etc.);
- young galleries made by the weevil should be sealed with mud and aluminium phosphate application (poisonous fumes);
- the grubs should be destroyed within the holes by injecting the above mentioned poisonous fumes.

To kill adult weevils inside the date palm, injection of insecticide into the trunk or fumigation could be practised. Phostoxin tablets are placed in infested trunks then sealed with gypsum or cement. No further injections into palms have been carried out in Saudi Arabia and Egypt since 1994, because they were found to be ineffective. There is no evidence from any country that chemical spraying/injecting has any effect on the rate of weevil infestations. Adult weevils can disperse about one km/day, which makes the process of chemical spraying a difficult one. Chemical treatment has proven to be positive only on cut and injured surfaces which, without this chemical treatment, will offer entry points to the weevil.

Regular Surveys

Infected and non infested areas need to be regularly surveyed, not only to detect and record new weevil infestations, but also to assess the health of uninfested plantations and the effectiveness of the adopted control measures. The frequency of these surveys depends on the life cycle of the weevil. Once a month during cold months, and twice a month during the early part of the warm season and summer time.

Pheromone Mass Trapping

The trapping and destroying of adults is a recent method of controlling the weevil. In the Middle East, where the attack by RPW is severe on date palm, pheromone-baited traps have been used for monitoring and for the reduction of the weevil population.

In 1993, a male produced aggregation pheromone was reported for *R. ferrugineus* and a pheromone-food trap was effective to capture large numbers of *R. ferrugineus*. Although males produce an aggregation pheromone that should attract equal numbers of males and females, the sex ratio of captured weevils is usually 3 - 4:1 in favour of females. It is worth mentioning that this mass trapping is successful only when combined with good sanitation and chemical control measures. It allows the reduction of the weevil population and the numbers of fl ying adults.

The use of pheromones have started in UAE, in Oman and the Kingdom of Saudi Arabia. Pheromone/food traps need to be placed where infestation is suspected/confi rmed at one (1) trap for each 100 metres. Traps need to be placed in the ground. According to Oehlschlager, the best trapping results are obtained if: - the pheromone lure contains pheromone and plant produced synergists; - food (such as date palm stem pieces, date fruit, sugar cane, bananas and apples) is kept wet by frequent addition of water; and - traps are shaded to keep them wet.

Use of Nematodes

The natural enemies of the weevil do not play a significant part in the control of its populations. However, in the Middle East the use of an entomopathogenic nematode (*H. indicus*) of *Heterorgabditis* species or *steinernema* sp. is being investigated. Third stage infective juveniles of the nematode in a symbiosis with *Xeonorhabdus* bacteria attack the weevil (grub stage only).

Desert Locust (Schistocerca Gregaria Forskal)

The desert locust occurs in all date growing areas of the Near East and North Africa and causes severe damage. Heavy migrations into date plantations are sporadic but may be devastating. The locust feeds on leaves and fruits of the date palm and may destroy the palm's canopy and leave the palm totally naked. Young locusts feed on younger plants and small offshoots.

Swarms of locusts are usually measured in terms of square miles and occur throughout the Old World date-growing areas. In fact, a swarm of 50 square miles represents about 10,000 tons of locusts.

To recover from a severe locust attack, a date plantation needs at least three years - under optimal growing conditions - to reconstitute its canopy. Within such period the fruit yield is of course heavily affected. Chemical control is effective if applied properly and well timed to kill locusts before they attack date palms. The use of aerial spraying on both ground and flying swarms of locusts (subspecies: gregaria) has been successful since 1959.

Rodents

Two types of rodents cause damage to date palm: The black rat (*Rattus rattus*) and the house mouse (*Mus musculus* L).

The black rat and the house mouse are usually in the field and storage area, and feed exclusively on date fruits. Besides damaging date fruits, rodents could also cause the following:

- establishment of underground galleries that threaten the traditional canal irrigation system and sometimes damage it;
- feeding on offshoot roots which affects their survival. It also feeds on roots of old palms causing them to fall down if feeding was only on one side of the palm and wind was severe;
- feeding on recently emerged infl orescences.

There is only one control measure, that is by using poison. A mixture of zinc phosphate at 30 to 50 g with 1 kg of millet fl our and 3 % of cooking oil. The paste is to be placed around the palms at the entry to the galleries. A chemical product "Finale" gave excellent results at the Eersbegin project (Namibia). It is a highly active anticoagulant bait at 0.025 g/kg as an active ingredient. The death of rodents takes 4 to 12 days. The chemical was recently used in both the Eersbegin and Naute date plantations (Namibia) with a sound success rate against *Mus musculus*.

Termites (Microcerotermes Diversus)

Termites usually feed on cellulose matter and the attack starts from the root zone and base of the offshoots by making vertical canals through it, or building soil-canals on it, allowing them to reach the stem. Where termites are found, they usually cause the death of newly planted offshoots. They may also make galleries in the trunks of weak palms and cause them to collapse.

Control measures could be started by removing and burning destroyed offshoots. In case of a slight attack, it is recommended to clean the offshoot of soil canals and spray it with a termite killer (Dursban or Hostathion). It is also advised to turn over the surrounding soil to about 50 cm deep in order to destroy these canals and treat them with a nematicide product (which will certainly kill all termite species).

Nematodes

Root-knot nematodes *(Meloidogyne spp.)* are widely distributed in date palm plantations, but the amount of damage caused to fruit bearing palms has not been determined. Nematodes are spread most readily by offshoots, which, if growing below the soil surface, may be infested while attached to the mother palm. Nurseries provide a second source of infestation of offshoots. Root-knot nematodes have such a wide range of cultivated and weed hosts that their control in date plantations has not been attempted. Dowson and Pansiot (1965) state that nematodes in the Old World date palm plantations do not appear to have been studied. It is possible that much of the unhealthy growth of palms, generally attributed to other causes, may be due to nematode attack.

Weeds

Weeds are plants that grow with date palms and act as competitors for food or serve as alternate hosts for insects and diseases. Numerous studies have established that weeds cause more damage than insects and fungi combined. They cause damage through reduction in yields, loss of nutrients and water, shading effect, increase in the cost of production and decrease in the quality of fruit and by acting as alternate hosts to other harmful organisms.

The most common weeds are: Haifa *(Imperata cylindrica),* Bermuda grass *(Cynodon dactylon), Cyperus spp., Chenopodium spp., Juncus* sp. and Johnson grass. Many other weeds of minor importance can be found in a date plantation.

A big obstacle in the adoption of effective weed control measures is the general lack of awareness of the impact of the damage caused by weeds. Various control attempts have been conducted to reduce weed damage. These include hoeing, ploughing, and chemical control. Improved weed management should be emphasized.

Date Palm Genome

In 2009, a team of researchers at the Weill Cornell Medical College in Qatar published a draft version of the date palm genome (Khalas variety).

Pomegranate

The pomegranate /ÈpRmharænht/, botanical name *Punica granatum*, is a fruit-bearing deciduous shrub or small tree growing between 5–8 metres (16–26 ft) tall.

The pomegranate is considered to have originated in Iran and has been cultivated since ancient times. Today, it is widely cultivated throughout the Mediterranean region of southern Europe, the Middle East and Caucasus region, northern Africa and tropical Africa, the Indian subcontinent, Central Asia and the drier parts of southeast Asia. Introduced into Latin America and California by Spanish settlers in 1769, pomegranate is also cultivated in parts of California and Arizona.

In the Northern Hemisphere, the fruit is typically in season from September to February. In the Southern Hemisphere, the pomegranate is in season from March to May.

The pomegranate has been mentioned in many ancient texts, notably in Babylonian texts, the Book of Exodus, the Homeric Hymns and the Quran. In recent years, it has become more common in the commercial markets of North America and the Western Hemisphere.

Pomegranates are used in cooking, baking, juices, smoothies and alcoholic beverages, such as martinis and wine.

The *Punica granatum* leaves are opposite or sub-opposite, glossy, narrow oblong, entire, 3–7 cm long and 2 cm broad. The flowers are bright red, 3 cm in diametre, with four to five petals (often more on cultivated plants). Some fruitless varieties are grown for the flowers alone. The edible fruit is a berry and is between a lemon and a grapefruit in size, 5–12 cm in diametre with a rounded hexagonal shape, and has thick reddish skin. The exact number of seeds in a pomegranate can vary from 200 to about 1400 seeds, contrary to some beliefs that all pomegranates have exactly the same number of seeds. Each seed has a surrounding water-laden pulp—the edible sarcotesta that forms from the seed coat—ranging in colour from white to deep red or purple. The seeds are "exarillate", i.e., unlike some other species in the Myrtales there is no aril. The sarcotesta of pomegranate seeds consists of epidermal cells derived from the integument, and there are no arils on these seeds. The seeds are embedded in a white, spongy, astringent membrane.

Cultivation

Figure: *Unripened Pomegranate fruit on a small tree in India*

Punica granatum is grown as a fruit crop plant, and as ornamental trees and shrubs in parks and gardens. Mature specimens can develop sculptural twisted bark multi-trunks and a distinctive overall form.

Pomegranates are drought-tolerant, and can be grown in dry areas with either a Mediterranean winter rainfall climate or in summer rainfall climates. In wetter areas, they can be prone to root decay from fungal diseases. They can be tolerant of moderate frost, down to about –12 °C (10 °F).

Insect pests of the pomegranate can include the pomegranate butterfly *Virachola isocrates* and the leaf-footed bug *Leptoglossus zonatus*. Pomegranate grows easily from seed, but is commonly propagated from 25–50 cm hardwood cuttings to avoid the genetic variation of seedlings. Air layering is also an option for propagation, but grafting fails.

Varieties

Punica granatum var. *nana* is a dwarf variety of *P. granatum* popularly planted as an ornamental plant in gardens and larger containers, and used as a bonsai specimen tree. It could well be a wild form with a distinct origin. It has gained the Royal Horticultural Society's Award of Garden Merit. The only other species in the genus *Punica* is the Socotran pomegranate (*Punica protopunica*), which is endemic to the island of Socotra. It differs in having pink (not red) flowers and smaller, less sweet fruit.

Cultivars

Punica granatum has more than 500 named cultivars, but the pomegranate evidently has considerable synonymy in which the same genotype is named differently across regions of the world.

Several characteristics between pomegranate genotypes vary for identification, consumer preference, preferred use, and marketing, the most important of which are fruit size, exocarp colour (ranging from yellow to purple, with pink and red most common), seed-coat colour (ranging from white to red), hardness of seed, maturity, juice content and its acidity, sweetness, and astringency.

Etymology and Terms for Pomegranate in Other Languages

The name *pomegranate* derives from medieval Latin *pômum* "apple" and *grânâtum* "seeded". This has influenced the common name for pomegranate in many languages (e.g. *granada* in Spanish, *Granatapfel* or *Grenadine* in German, *grenade* in French, *granatäpple* in Swedish, *pomogranà* in Venetian). *Mâlum grânâtus*, using the classical Latin word for apple, gives rise to the Italian name *melograno*, or less commonly *melagrana*.

Perhaps stemming from the old French word for the fruit, *pomme-grenade*, the pomegranate was known in early English as "apple of

Grenada"—a term which today survives only in heraldic blazons. This is a folk etymology, confusing Latin *granatus* with the name of the Spanish city of Granada, which derives from Arabic.

The genus name *Punica* refers to the Phoenicians, who were active in broadening its cultivation, partly for religious reasons.

Garnet comes from Old French *grenat* by metathesis, from Medieval Latin *granatum*, here used in a different meaning: "of a dark red colour". This meaning perhaps originated from *pomum granatum* because of the colour of pomegranate pulp, or from *granum* in the sense of "red dye, cochineal".

The French term *grenade* for pomegranate has given its name to the military grenade. Soldiers commented on the similar shape of early grenades and the name entered common usage.

While most European languages have cognate names for the fruit, stemming from Latin *granatum*, exceptions are the Armenian term *nur*, Albanian term *shega*, Bulgarian *nar* and the Portuguese term *romã* which is derived from Arabic *ruman*, and has cognates in other Semitic languages (e.g. Hebrew *rimmon*) and Ancient Egyptian *rmn*.

Medicinal Values / Uses

The juice of pomegranate is used for pharmaceutical purposes. It is used for treating dyspepsia and is considered beneficial in treating leprosy. Because of the tannin content of the bark, leaves, immature fruits and fruit rind are given as astringent to halt diarrhea, dysentery and hemorrhages. A decoction of the flowers is gargled to relieve oral and throat inflammation.

Major Growing Areas

Pomegranate is adapted to a wide range of climate conditions. It does well under desert conditions and produce high quality fruits when grown in loamy soil especially in locations with cool winter and a dry summer. In Sri Lanka most suitable districts for cultivation are Hambantota, Puttalum, Manar, Anuradapura, Jaffna, Monaragala.

Recommended Varieties

Local cultivars

Crop Management

Nature of Cultivation

Scattered; home gardens; also scattered in the Wet zone (less productive)

Pest and Disease Incidence

Pests: Almost all Pomegranate varieties are found to be susceptible to the Pomegranate butterfly (Virachola isocrates) which is the most important pest in Sri Lanka. The caterpillar of the butterfly enters the fruits and feed on it.

Diseases: There are several fungus diseases which are economically important in Pomegranate cultivation and are detailed below.

Fruit Rot

i. Anthracnose fruit rot - Caused by Sphaceloma punicae

ii. Aspergillious fruit rot - Caused by Aspergillus spp.

iii. Penicillum fruit rot - Caused by Penicillium spp.

Leaf Spots

Causal organisms are Colletotrichum spp., Phomopsis spp., Cercospora punicae.

Die back: Caused by either Pleuroplaconema spp. or Centhosphora phyllostica Fruit cracking Cause is not fully understood. However it is suspected to be a physiological disorder.

Harvesting & Post-harvest Technology

Fresh fruits of Pomegranate could be used for enjoying at the table. In some countries such as Iran, the juice is a very popular beverage. In future, fruit juice extraction can be introduced as a cottage industry in Sri Lanka. Also, extracted juice may be concentrated and frozen for future use. It may be made in to a thick syrup for use as a sauce. It is also often converted into wine.

Pistachio (*Pistachia Vera*)

Pistachio (*Pistachia vera*) does well especially in the mid to high desert areas. The winters in the low desert (Phoenix or Palm Springs) may not be cold enough; they need about 1000 hours of winter chilling to produce nuts. Pistachios need both a male and female trees for pollination. The 'Peters' is the most commonly planted male tree; the 'Kerman' the best female variety. They grow to 25 to 30 feet tall and take 5 to 8 years to begin producing nuts. They prefer well-drained soil and lots of summer sun. Deep, very infrequent watering after the trees are established. Watering too often will produce root rot. To harvest the nuts, use a tarp on the ground and shake the tree in early autumn.

The pistachio, *Pistacia vera,* a member of the cashew family, is a small tree originally from Central Asia and the Middle East. Pistachio trees can be found in regions of Iran, Syria, Lebanon, Turkey, Greece, Tunisia, Kyrgyzstan, Tajikistan, Turkmenistan, India, Pakistan, Egypt, Italy (Sicily), Uzbekistan, Afghanistan (especially in the provinces of Samangan and Badghis), and the United States, specifically in California. The tree produces seeds that are widely consumed as food.

However, it is only native to the eastern Mediterranean (Cyprus and Turkey to Israel and Syria) and Central Asia (Turkmenistan, Uzbekistan, Afghanistan, Tajikistan, and Kyrgyzstan).

Pistacia vera often is confused with other species in the genus *Pistacia* that are also known as pistachio. These species can be distinguished from *P. vera* by their geographic distributions (in the wild), and their seeds which are much smaller and have a shell that is not hard.

Botany

Habitat: Pistachio is a desert plant, and is highly tolerant of saline soil. It has been reported to grow well when irrigated with water having 3,000–4,000 ppm of soluble salts. Pistachio trees are fairly hardy in the right conditions, and can survive temperatures ranging between –10 °C (14 °F) in winter and 48 °C (118 °F) in summer. They need a sunny position and well-drained soil. Pistachio trees do poorly in conditions of high humidity, and are susceptible to root rot in winter if they get too much water and the soil is not sufficiently free-draining. Long, hot summers are required for proper ripening of the fruit.

The Jylgyndy Forest Reserve, a preserve protecting the native habitat of *Pistacia vera* groves, is located in the Nooken District of Jalal-Abad Province of Kyrgyzstan.

Characteristics

Figure: *Pistachios in shell*

The bush grows up to 10 metres (33 ft) tall. It has deciduous pinnate leaves 10–20 centimetres (4–8 inches) long. The plants are dioecious, with separate male and female trees. The flowers are apetalous and unisexual, and borne in panicles.

The fruit is a drupe, containing an elongated seed, which is the edible portion. The seed, commonly thought of as a nut, is a culinary nut, not a botanical nut. The fruit has a hard, whitish exterior shell. The seed has a mauvish skin and light green flesh, with a distinctive flavor. When the fruit ripens, the shell changes from green to an autumnal yellow/red, and abruptly splits part way open. This is known as dehiscence, and happens with an audible pop. The splitting open is a trait that has been selected by humans. Commercial cultivars vary in how consistently they split open.

Each pistachio tree averages around 50 kilograms (110 lb) of seeds, or around 50,000, every two years.

The shell of the pistachio is naturally a beige colour, but it is sometimes dyed red or green in commercial pistachios. Originally, dye was applied by importers to hide stains on the shells caused when the seeds were picked by hand. Most pistachios are now picked by machine and the shells remain unstained, making dyeing unnecessary except to meet ingrained consumer expectations. Roasted pistachio seeds can be artificially turned red if they are marinated prior to roasting in a salt and strawberry marinade, or salt and citrus salts.

Like other members of the Anacardiaceae family (which includes poison ivy, sumac, mango, and cashew), pistachios contain urushiol, an irritant that can cause allergic reactions.

Cultivation

Iran, the United States and Turkey are the major producers of pistachios. The trees are planted in orchards, and take approximately seven to ten years to reach significant production. Production is alternate bearing or biennial bearing, meaning the harvest is heavier in alternate years. Peak production is reached at approximately 20 years. Trees are usually pruned to size to make the harvest easier. One male tree produces enough pollen for eight to twelve drupe-bearing females. Harvesting in the United States and in Greece is often accomplished by using shaking equipment to shake the drupes off the tree. After hulling and drying, pistachios are sorted according to open mouth and closed mouth shell. Sun drying has been found to be the best method of drying. Then they are roasted or processed by special machines to produce pistachio kernels.

Pistachio trees are vulnerable to a wide variety of diseases. Among these is infection by the fungus *Botryosphaeria*, which causes panicle and shoot blight (*i.e.,* kills flowers and young shoots), and can damage entire pistachio orchards.

In California, almost all female pistachio trees are the cultivar "Kerman". A scion from a mature female Kerman is grafted onto a one-year-old rootstock. Male pistachios may be a different variety.

In 2013, the City Council of Avenal, California self-proclaimed Avenal to be the Pistachio Capital of the World, reflecting the importance of the pistachio industry to the city's economy.

In Greece, the cultivated type of pistachios is different. It has an almost-white shell, a sweet taste, a red-green kernel and a little bit more close mouth shell than "Kerman" variety. Most of the production in Greece comes from the island of Aegina and the region of Thessaly - Almyros.

Bulk container shipments of pistachio kernels are prone to self-heating and spontaneous combustion because of their high fat and low water content.

Consumption

The kernels are often eaten whole, either fresh or roasted and salted, and are also used in pistachio ice cream, kulfi, spumoni, historically in Neopolitan ice cream, pistachio butter, pistachio paste and confections such as baklava, pistachio chocolate, pistachio halva, pistachio lokum or biscotti and cold cuts such as mortadella. Americans make pistachio salad, which includes fresh pistachios or pistachio pudding, whipped cream, and canned fruit.

In July 2003, the Food and Drug Administration (FDA) approved the first qualified health claim specific to seeds lowering the risk of heart disease: "Scientific evidence suggests but does not prove that eating 1.5 ounces (42.5 g) per day of most nuts, such as pistachios, as part of a diet low in saturated fat and cholesterol may reduce the risk of heart disease".

China is the top pistachio consumer worldwide, with annual consumption of 80,000 tons, while the United States consumes 45,000 tons. Russia (with consumption of 15,000 tons) and India (with consumption of 10,000 tons) are in the third and fourth places.

Health Benefits

In research at Pennsylvania State University, pistachios in particular significantly reduced levels of low-density lipoprotein (LDL cholesterol) while increasing antioxidant levels in the serum of volunteers. In rats, consumption of pistachios as 20% of daily caloric intake increased beneficial high-density lipoprotein (HDL cholesterol) without lowering LDL cholesterol, and while reducing LDL oxidation.

Human studies have shown that 32–63 grams (1.1–2.2 oz) per day of pistachio seeds can significantly elevate plasma levels of lutein, alpha-carotene, beta-carotene, and gamma-tocopherol.

In December 2008, James Painter, a behavioural eating expert, professor and chair of School of Family and Consumer Sciences at Eastern Illinois University, described the Pistachio Principle. The Pistachio Principle describes methods of "fooling" one's body into eating less. One example used is that the act of shelling and eating pistachios one by one slows one's consumption, allowing one to feel full faster after having eaten less. Contrarily, pistachios contain a somewhat high amount of palmitic fatty acid, a fatty acid that is associated with an increase in appetite and cardiovascular disease by the World Health Organization.

The fat profile of pistachios is roughly 14% saturated fat, 54% monounsaturated fat and 32% polyunsaturated fat. A 1-cup serving of pistachio seeds contains 6.143 g (94.80 gr) of saturated palmitic fatty acid, only 0.585 g (9.03 gr) of stearic fatty acid and trace amounts of arachidic and behenic saturated fatty acids.

Preliminary studies show pistachios have antibacterial activity against the intestinal pathogens *Escherichia coli* and *Listeria monocytogenes*. Further studies are needed to establish if dietary pistachio intake offers any protection against foodborne infection.

Toxin and Safety Concerns

As with other tree seeds, aflatoxin is a toxin found in poorly harvested or processed pistachios. Aflatoxins are potent carcinogenic chemicals produced by molds such as *Aspergillus flavus* and *Aspergillus parasiticus*. The mold contamination may occur from soil, poor storage, and spread by pests. High levels of mold growth typically appear as gray to black filament-like growth. It is unsafe to eat mold-infected and aflatoxin-contaminated pistachios. Aflatoxin contamination is a frequent risk, particularly in warmer and humid environments. Food contaminated with aflatoxins has been found as the cause of frequent outbreaks of acute illnesses in parts of the world. In some cases, such as Kenya, this has led to several deaths.

Pistachio shells typically split naturally prior to harvest, with a hull covering the intact seeds. The hull protects the kernel from invasion by molds and insects. But this hull protection can be damaged in the orchard by poor orchard management practices, by birds, or after harvest, which makes it much easier for pistachios to be exposed to contamination. Some pistachios undergo so-called "early split," wherein both the hull and the shell split. Damage or early splits can lead to aflatoxin contamination.

In some cases, a harvest may be treated to keep contamination below strict food safety thresholds; in other cases, an entire batch of pistachios must be destroyed because of aflatoxin contamination. In September 1997, the European Union placed its first ban on pistachio imports from Iran due to high levels of aflatoxin. The ban was lifted in December 1997 after Iran introduced and improved food safety inspections and product quality.

Recycling the Shells

Empty pistachio shells can be recycled in several ways. If unsalted, the shells need neither washing nor drying before reuse but washing is simple if this is not the case. Practical uses include as a fire starter, just as crumpled paper is used as kindling; to line the bottom of pots containing houseplants, for drainage and retention of soil for up to two years; as a mulch for shrubs and plants that require acid soils; as a medium for orchids; and as an addition to a compost pile designed for wood items that take longer to decompose than leafy materials, taking up to a year for pistachio shells to decompose unless soil is added to the mix. Shells from salted pistachios can also be placed around the base of plants to deter slugs and snails. Craft uses for the shells include Christmas tree ornaments, jewelry, mosaics, and rattles. Research indicates that pistachio shells may be helpful in cleaning up pollution created by mercury emissions.

Almonds (*Prunus Dulcis*)

Almonds (*Prunus dulcis*) love hot, dry summer weather. The recommended variety for home gardens is the semi-dwarf 'All-In-One' which self-pollinates and grows best in reasonably fast-draining soil. The large, soft-shelled nuts look like clusters of small green peaches until late summer when the outer coat peels back to reveal the almond in its shell. The 'All-In-One' almond tree reaches 15 feet in height.

The other almond trees that grow well in the desert are full-size, reaching 20 to 30 feet tall, and require two varieties to pollinate and produce nuts. If you select the 'Nonpareil' or the 'Ne Plus Ultra', they will pollinate each other. The 'All-in-One' will also pollinate both of these trees. To harvest the nuts, place a tarp on the ground and shake the trees. Dry fallen nuts in the sun for a day or two until the nuts inside the shell rattle.

The almond is a species of tree native to the Middle East and South Asia. "Almond" is also the name of the edible and widely cultivated seed of this tree. Within the genus *Prunus*, it is classified with the peach in the subgenus *Amygdalus*, distinguished from the other subgenera by the corrugated shell (endocarp) surrounding the seed.

The fruit of the almond is a drupe, consisting of an outer hull and a hard shell with the seed (which is not a true nut) inside. Shelling almonds refers to removing the shell to reveal the seed. Almonds are sold shelled (i.e., after the shells are removed), or unshelled (i.e., with the shells still attached). Blanched almonds are shelled almonds that have been treated with hot water to soften the seedcoat, which is then removed to reveal the white embryo.

Description

Tree: The almond is a deciduous tree, growing 4–10 m (13–33 ft) in height, with a trunk of up to 30 cm (12 in) in diametre. The young twigs are green at first, becoming purplish where exposed to sunlight, then grey in their second year. The leaves are 3–5 inches long, with a serrated margin and a 2.5 cm (1 in) petiole. The flowers are white to pale pink, 3–5 cm (1–2 in) diametre with five petals, produced singly or in pairs and appearing before the leaves in early spring.

Almonds begin bearing an economic crop in the third year after planting. Trees reach full bearing five to six years after planting. The fruit matures in the autumn, 7–8 months after flowering.

Drupe: The almond fruit measures 3.5–6 cm (1–2 in) long. In botanical terms, it is not a nut, but a drupe. The outer covering or exocarp, fleshy in other members of *Prunus* such as the plum and cherry, is instead a thick, leathery, grey-green coat (with a downy exterior), called the hull. Inside the hull is a reticulated, hard, woody shell (like the outside of a peach pit) called the endocarp. Inside the shell is the edible seed, commonly called a nut. Generally, one seed is present, but occasionally two occur.

Production

The world produced 2.00 million tonnes of almonds in 2011 according to Food and Agriculture Organization, with United States the largest producer at 0.73 million tonnes. The apparent 50% decrease in production by the United States led to a calculated percent of world production decrease from 56% to 36%; however, a 2013 news article indicated that the United States produced at least 80% of the world's supply.

Spain has one of the most diverse commercial cultivars of almonds. It is grown in Spain's Catalonia, Valencia, Murcia, the Balearic Islands,

Andalusia, and Aragón regions. In Greece, most of the production comes from the region of Magnesia at the area of Almyros. The most cultivating types of almonds in Greece are ferragnes and Texas (mission) which are known for their sweet taste and premium quality. Because of its quality, it is used as a luxury nut. In Turkey, most of the production comes from the Aegean, Marmara, and Mediterranean regions.

In the United States, production is concentrated in California, with almonds being California's third leading agricultural product and its top agricultural export in 2008. and 100% of the U.S. commercial supply. The United States is the dominant supplier of almonds. In 2011, the country exported about 637,000 metric tons, valued at US$2.8 billion. Almonds were mostly exported as shelled almonds (70%), with the remainder being either unshelled or processed.

Australia is the largest almond production region in the southern hemisphere. In 2013, Australia contributed to 5.9% of the world almond supply. Most of the almond orchards are located in the states of New South Wales, Victoria and South Australia.

In 2013 and 2014, environmental problems in California affected the almond supply, contributing to higher almond prices worldwide.

Pollination

The pollination of California's almonds is the largest annual managed pollination event in the world, with close to one million hives (nearly half of all beehives in the USA) being trucked in February to the almond groves. Much of the pollination is managed by pollination brokers, who contract with migratory beekeepers from at least 49 states for the event. This business has been heavily impacted by colony collapse disorder, causing nationwide shortages of honey bees and increasing the price of insect pollination.

To alleviate almond growers from the rising cost of insect pollination, researchers at the Agricultural Research Service (ARS) have developed a new line of self-pollinating almond trees. Self-pollinating almond trees, such as the 'Tuono', have been around for a while, but their harvest is not as desirable as the insect-pollinated California 'Nonpareil' almond tree. The 'Nonpareil' tree produces large, smooth almonds and offers 60–65% edible kernel per nut. The 'Tuono', however, has thicker, hairier shells and offers only 32% of edible kernel per nut, but having a thick shell has advantages. The 'Tuono's' shell protects the nut from threatening pests such as the navel orangeworm. ARS researchers have managed to cross-breed the pest-resistant 'Tuono' tree with the 'Nonpareil', resulting in hybridized varieties of almond trees that are self-pollinated and maintain a high nut quality. The new, self-pollinating hybrids possess quality skin colour, flavor, and oil content, and reduce almond growers' dependency on insect pollination.

Diseases

Almond trees can be attacked by an array of damaging organisms, including insects, fungal pathogens, plant viruses and bacteria. Symptoms can vary depending on the organism attacking the plant.

Insect pest management in almonds uses integrated techniques that in many cases allow growers to avoid chemical treatments entirely. Insects are monitored with pheromone traps, sticky traps, and egg traps. Day-degree models define insect life cycles and can identify the potential for crop damage based on the coincidence of pests with susceptible crop stages. Improved control timing allows intervention with selective, short-residual spray materials that are less harmful to beneficial insects. Cultural practices to minimize pest damage are employed, and, releases of beneficial insects may further reduce crop damage.

San jose scale (Quadraspidiotus perniciosus) was traditionally controlled with annual dormant applications of oil and organophosphate insecticide.

Now a dormant spur sampling can be used to gauge the need for a dormant oil spray and dormant oil alone applied in early to mid-January can provide control of young scales. Male scale flights during the growing season can be easily monitored with pheromone traps. These traps also catch parasites and can be used to judge if they are present in the orchard based on incidental trap catches.

Aphytis spp. Or Encarsia perniciosi scale parasitoids can be easily distinguished from the male scales in the traps. A day-degree (Do) phenology model can predict crawler emergence (beginning 405 Do after the male flight) and can help time a narrow range oil spray to be most effective at 600 to 700 Do if the parasitoids need help in controlling the scale. This method can be employed if significant scale populations are noted.

Navel orangeworm (Ameylois transitella) is the major worm pest of almonds. It overwinters in mummy nuts left on the tree after harvest and the first generation in the spring will continue to feed and increase on the remaining mummy nuts. Navel orangeworm infests the new crop once hullsplit begins in July. Winter orchard sanitation to destroy mummy nuts is the most effective control for this pest. Removing the mummy nuts from the trees in the winter and chopping them when the orchard floor is mowed in early spring destroys a large percentage of the overwintering worm population and prevents a spring population in- crease. This cultural control is a much more effective than parasite releases or in-season sprays timed at early hull split.

During the growing season, egg traps baited with ground almond meal moistened with almond oil are used to attract and monitor egg-laying activity and to identify a biofix in the spring. Using egg trap data and the navel orangeworm phenology model, generations can be predicted so that an early harvest can be accomplished in August perhaps before the third generation eggs are laid on the nuts.

Traditionally, almond harvest began about 1 Sept. after the nuts were fully split and dry on the tree. Since the navel orangeworm only lays eggs on nuts in the tree, an early harvest when 100 % of nuts are split but not fully dry can help avoid this egg laying, and, crop damage can be greatly reduced.

Early harvest has been widely adopted and when followed by rapid hulling and processing, or, fumigation of stock-piled nuts waiting for hulling, worm damage is minimized.

The parasitic wasp, Goniozus legneri, can help manage the navel orangeworm population most effectively when it is reared and released early in the growing season. These tiny wasps sting the worm larvae paralyzing it in preparation for egg laying. An egg is laid and the external parasite develops by feeding on the paralyzed larva. At hullsplit, an organophosphate insecticide spray (an old standard control) or Bacillus Thuringiensis sprays may only be needed if these effective cultural practices and biocontrols have not been thoroughly implemented.

Peach twig borer (Anarsia lineatella), generally feeds in new shoots and can damage new growth on first and second year trees. However, it can also feed on the nuts causing considerable damage. It is effectively controlled with organophosphate dormant sprays aimed at peach twig borer larvae that overwinter in hibernacula incrotches of small twigs. These dormant organophosphate sprays are be- ing phased out since small amounts of pesticide residue have been found in rivers following heavy rains.

A widely adopted alternative is the use of Bacillus thuringiensis sprays applied following full bloom and petal fall when the tiny larvae emerge from the hibernacula and feed on the new green growth. Pheromone traps and a phenology model can help time sprays in May if needed on young trees or in July if the pest population coincides with hullsplit in bearing orchards.

Ants damage almonds primarily by feeding on the kernels while the nuts are drying on the ground during harvest. Ant control must be done earlier in the season before damage is observed. The southern fire ant (Solenopsis xyloni) and the pavement ant (Tetramorium caespitum) are the two damaging ant species that need to be controlled. Other ant species are not harmful or may actually be beneficial such as the gray ant (Formica aerata) which is an excellent peach twig borer predator. Whenever ants are observed, they need to be identified before control measures are implemented.

Field monitoring can be done in June with open vials baited with pieces of hot dog or almond kernels and placed on the ground at intervals throughout the orchard. If significant numbers of damaging ants are detected a selective baiting programme can be implemented in the appropriate areas to reduce their numbers before harvest.

Webspinning spider mite management is more effective when a presence/absence sequential sampling scheme is used to help make treatment decisions. This scheme involves sampling weekly a minimum of 15 leaves per tree on five trees in mite hot spots and quickly rating each leaf as to whether mites and their predators are present or absent. Counts are compared to a treatment decision chart that helps determine whether or not a treatment is warranted. The web spinning spider mites, pacific spider mite (Tetranychus pacificus) and twospotted spider mite (Tetranychus urticae), can build up quickly to high populations that cause defoliation.

Monitoring the population weekly with this sampling scheme prevents surprises and helps determine whether or not a mite treatment is needed. Multiple tactics are used for mite management. These include management to favour mite predators, good irrigation, dust control, and early treatment of mite "hot spots" with low rates of selective acaricides. Management of the beneficial western predatory mite (Galendromus occidentalis) may include supplemental release of these predator mites into the orchard. Orchard wide acaricide application is used only as a last resort.

Disease Management

Disease control has traditionally been based on prevention with fungicides. Brown rot (Monilinia laxa), a disease affecting almond flowers, is a good example of this approach. Control of this disease is simply a matter of applying one fungicide spray at pink bud as a bloom protectant. Available materials are generally very effective on brown rot, and, once treated, the disease is prevented. There are a number of fungicides available for use in almonds that have varying efficacy against the range of almond diseases.

New selective fungicides available today often have a shorter residual activity than older fungicides. This requires better knowledge of the target disease and improved application timing. These new fungicides may also have a single site mode of action, which requires that a resistance management strategy be in place. In wet springs when multiple fungicide applications are required, products from different classes must be selected and used in rotation.

Three diseases of almond, shot holefungus (Wilsonomyces carpophilus), almond scab (Cladosporium carpophilum), and anthracnose (Colletotrichum acutatum) are significant enough that management strategies going beyond fungicide application have been developed.

Figs (*Ficus Carica*)

Figs (*Ficus carica*) are big leaf trees that grow fast to 15 to 30 feet in height. They love the heat and do well when planted near a south-facing wall — but not too close. Eventually the tree trunk becomes quite large. Most varieties produce 2 crops a year and, for home garden use, do not need another fig tree to for pollination. The best varieties for the desert are: 'Black Mission', 'Kadota' and 'Brown Turkey'.

Pecans (*Carya illinoisis*) round out the mix of nuts that can be grown in our Southwestern desert climate. Most pecans will produce a light crop of nuts with just a single tree. For a larger harvest, plant two of these trees. But be forewarned — these are large trees, growing to 70 feet and, unlike the pistachios or almonds, pecans require regular watering if you want a good crop of nuts. The best varieties for the desert are the 'Western Shley', 'Mahan' and 'Wichita'. If your soil has a high salt content, you should probably choose another tree to plant. The nuts will fall from the tree when ripe.

Apricot trees (*Prunus family*) reach 15 to 20 feet in height and have pink or white blooms in Spring. Varieties that do well in the desert are: 'Early Gold', 'Blenheim', 'Royal', 'Chinese', 'Tilton', 'Floragold' (a dwarf variety), and 'Newcastle'. Most of these are self-pollinating and need some winter chill.

Plum

Plum trees (*Prunus*) reach 10 to 15 feet in height and will need a winter chill period to produce abundant fruit. Among the best varieties for our hot, dry climate are two self-pollinators: 'Beauty' and 'Santa Rosa'. The 'Satsuma'. 'Burbank', 'Howard Miracle', 'Mariposa' and 'Friar' can be pollinated by the 'Santa Rosa'. There are, of course, the ornamental plums, but why grow them when you can grow fruit bearing trees!

Nectarine trees (*Prunus persica nucipersica*) need to be pruned back severely every year because the fruit grows only on the first year growth. Even then you are likely to have a bumper crop annually. Plant these self-pollinating varieties: 'Goldmine', 'Gower', 'Stanwick', and 'Le Grand' and you can feed the whole neighbourhood!

Peach trees (*Prunus persica*) as well as nectarine trees grow to about 25 feet high, if left unpruned. Pruning is recommended to keep tree height to under 12 feet. They will start producing fruit in about 3 or 4 years and you can place 2 or 3 varieties in one hole when you plant. Some varieties that do well in the desert are: 'Desert Gold', 'Early Elberta', 'Bonita', and 'Rio Grande'. The following are dwarf trees: 'Bonanza II', 'Southern Sweet', and 'Southern Flame'.

6

Under and Un-utilized Fruits of Tropics and Sub-tropics

Underutilized crops have the potential to provide great social and nutritional benefits to the rural poor as well as the global community at large, and it is increasingly recommended that they are identified, researched and promoted in much the same ways and levels of resourcing as the world's main staple food plants. Underutilized crops are defined as 'species' with under-exploited potential for contributing to food security, health (nutritional/medicinal), income generation and environmental services. South-East Asia and many of the islands throughout the Pacific have tradi-tional tropical and subtropical fruits. These fruits are not extensively cultivated, and consumption and trade are limited. However, many of these are of considerable economic importance in their respective regional markets. Nutritional analysis of the region's underutilized species is now being seen as an important component of plant genetic resource project. Many of these species are well- used by local communities and their nutritional and medicinal benefits are well known. The diversity of under-utilized tropical and subtropical fruits that exist in this region is highly significant and could equip the communities well with by opening up high value-niche market.

Ziziphus Mauritiana

Ziziphus mauritiana, also known as Ber, Chinee Apple, Jujube, Indian plum and Masau is a tropical fruit tree species belonging to the family Rhamnaceae.

Z. mauritiana is a spiny, evergreen shrub or small tree up to 15 m high, with trunk 40 cm or more in diametre; spreading crown; stipular spines and many drooping branches. The fruit is of variable shape and size. It can be oval, obovate, oblong or round, and that can be 1-2.5 in (2.5-6.25 cm) long, depending on the variety. The flesh is white and crisp. When slightly underipe, this fruit is a bit juicy and has a pleasant aroma. The fruit's skin is smooth, glossy, thin but tight.

The species is believed to have originated in Indo-Malaysian region of South-East Asia. It is now widely naturalised throughout the Old World tropics from Southern Africa through the Middle East to the Indian Subcontinent and China, Indomalaya, and into Australasia and the Pacific Islands. It can form dense stands and become invasive in some areas, including Fiji and Australia and has become a serious environmental weed in Northern Australia. It is a fast growing tree with a medium lifespan, that can quickly reach up to 10–40 ft (3 to 12 m) tall.

Botany

Figure: Z. mauritiana *tree*

Z. mauritiana is a medium sized tree that grows vigorously and has a rapidly developing taproot, a necessary adaptation to drought conditions. The species varies widely in height, from a bushy shrub 1.5 to 2 m tall, to a tree 10 to 12 m tall with a trunk diametre of about 30 cm. *Z. mauritiana* may be erect or wide-spreading, with gracefully drooping thorny branches, zigzag branchlets, thornless or set with short, sharp straight or hooked spines.

The leaves are alternate, ovate or oblong elliptic with rounded apex, with 3 depressed longitudinal veins at the base. The leaves are about 2.5 to 3.2 cm long and 1.8 to 3.8 cm wide having fine tooth at margin. It is dark-green and glossy on the upper side and pubescent and pale-green to grey-green on the lower side. Depending on the climate, the foliage of the *Z. mauritiana* may be evergreen or deciduous.

The flowers are tiny, yellow, 5-petalled and are usually in twos and threes in the leaf axils. Flowers are white or greenish white and the fruits are orange to brown, 2–3 cm long, with edible white pulp surrounding a 2-locular pyrene.

This quick growing tree starts producing fruits within three years. The fruit is a soft, juicy, drupe that is 2.5 cm diametre though with sophisticated cultivation the fruit has of size 6.25 cm long and 4.5 cm wide. The form may be oval, obovate, round or oblong; the skin smooth or rough, glossy, thin but tough. The fruit ripen at different times even on a single tree. Fruits are first green, turning yellow as they ripen. The fully mature fruit is entirely red, soft, juicy with wrinkled skin and has pleasant aroma. The ripe fruit is sweet and sour in taste. Both flesh texture and taste are reminiscent of apples. When under ripe the flesh is white and crispy, acid to subacid to sweet in taste. Fully ripe fruits are less crisp and somewhat mealy; overripe fruits are wrinkled, the flesh buff-coloured, soft, spongy and musky. At first the aroma is apple like and pleasant but it becomes peculiarly musky when overripe. There is a single, hard, oval or oblate, rough central stone which contains 2 elliptic, brown seeds, 1/4 in (6mm) long.

Ecology

Z. mauritiana is hardy tree that copes with extreme temperatures and thrives under rather dry conditions with an annual rainfall of 6 to 88.5 in (15–225 cm). In Fiji, sometimes naturalised Ber trees grow along roadsides and in agricultural land, usually near sea level but occasionally up to an elevation of about 600 m. It also grows well on laterite, medium black soils with good drainage, or sandy, gravelly, alluvial soil of dry river-beds where it is vigorously spontaneous. In Australia, this species grows on a wide variety of soil types, including cracking clays, solodic soils and deep alluvials, in the tropics and subtropics where the average annual rainfall is in the range 470-1200mm. In the drier parts of this range, it grows best in riparian zones. Commercial cultivation usually extends up to 1000 m. Beyond this elevation trees do not perform well, and cultivation becomes less economical.

The tree has a high tolerance to both water-logging and drought and can grow where annual rainfall ranges from 125 to 2,225 mm, but is more widespread in areas with an annual rainfall of 300 to 500 mm. In China and India, wild trees are found up to an elevation of 5,400 ft (1,650 m). In India, the minimum shade temperature for survival is

7–13° and the maximum temperature is 50 °C. Studies report that this species flourishes in alkaline soils with a pH as high as 9.2. However, deep sandy loam to loamy soils with neutral or slightly alkaline pH are considered optimum for growth. In India, the tree grows best on sandy loam, neutral or slightly alkaline.

In India, there are 90 or more cultivars depending on the habit of the tree, leaf shape, fruit form, size, colour, flavor, keeping quality, and fruiting season. Among the important cultivars, eleven are described in the encyclopaedic Wealth of India: 'Banarasi (or Banarsi) Pewandi', 'Dandan', 'Kaithli' ('Patham'), 'Muria Mahrara', 'Narikelee', 'Nazuk', 'Sanauri 1', 'Sanauri 5', 'Thornless' and 'Umran' ('Umri'). The skin of most is smooth and greenish-yellow to yellow.

Reproductive Biology

Some cultivars attain anthesis early in the morning, others do so later in the day. The flowers are protandrous. Hence, fruit set depends on cross-pollination by insects attracted by the fragrance and nectar. Pollen of the Indian jujube is thick and heavy. It is not airborne but is transferred from flower to flower by honeybees. The flowers are pollinated by ants and other insects, and in the wild state the trees do not set fruits by self-pollination. Ber propagates by seeds, seedlings, direct sowing, root suckers as well as by cuttings. Ber seeds are spread by birds, native animals, stock, feral pigs and humans who eat the fruit and expel the seeds. Seeds may remain viable for 2½ years but the rate of germination declines with age. Cross-incompatibility occurs, and cultivars have to be matched for good fruit set; some cultivars produce good crops parthenocarpically.

Propagation

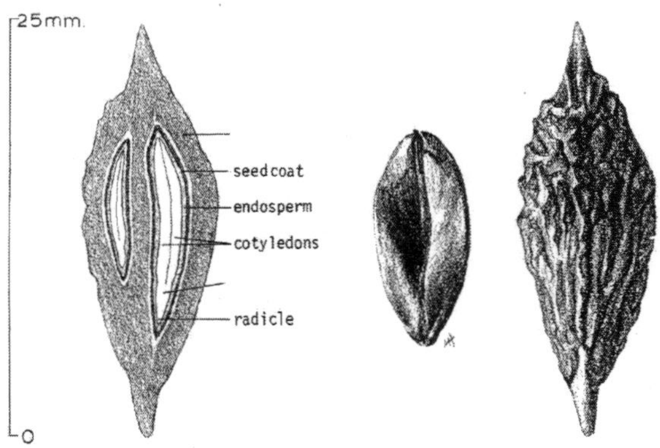

Figure: *Internal structure of Common jujube seed*

Z. mauritiana is one of the two *Ziziphus* species that have considerable horticulture importance, the other being Chinese jujube

(*Z. zizyphus*). Indian jujube is more tropical whereas Chinese jujube is a more cold hardy species.

Propagation is most commonly from seed, where pretreatment is beneficial. Storage of the seed for 4 months to let it after-ripen improves germination. The hard stone restricts germination and cracking the shell or extraction of seeds hastens germination. Without pretreatment the seeds normally germinate within six weeks whereas extracted seeds only need one week to germinate

Seedlings to be used as rootstock can be raised from seed. Several studies indicate that germination can be improved by soaking seeds in sulfuric acid. Germination time can also be shortened to 7 days by carefully cracking the endocarp. Ber seedlings do not tolerate transplanting, therefore the best alternatives are to sow the seeds directly in the field or to use polythene tubes placed in the nursery bed. Seedlings are ready for budding in 3 to 4 months. In addition, seedlings from the wild cultivars can be converted into improved cultivars by top-working and grafting. Nurseries are used for large scale seedling multiplication and graft production. The seedlings should also be given full light. The seedlings may need as long as 15 months in the nursery before planting in the field.

Scientists in India have standardised propagation techniques for Ber establishment. Budding is the easiest method of vegetative propagation used for improved cultivars. Different types of budding techniques have been utilised with ring-budding and shield-budding being the most successful. Wild varieties of ber are usually used as the root-stock. The most common being Z. rotundifolia in India and Z. spina-christi in Africa.

Season and Harvesting

Plants are capable of seed production once they reach a height of about 1 metre. Wild-growing plants in northern Australia may take 8 years to reach this size. In Australia, plants growing under natural conditions are capable of producing seeds once they reach a height of about 1m. Plants between 1 and 2m high produce, on average, less than five fruits per season. Large plants (>5m high) can produce 5000 or more fruits in a single season.

In India, some types ripen as early as October, others from mid-February to mid-March, others in March, or mid-March, to the end of April. In the Assiut Governorate, there are 2 crops a year, the main in early spring, the second in the fall. In India the trees flower in July to October and fruits are formed soon after. In February–March the fruits are mature and in some places a second crop is produced in the fall. Pickings are done by hand from ladders and about 110 lbs (50 kg)is harvested per day. The fruits remaining on the tree are shaken down. Only fully mature fruits are picked directly from the tree. They are transported in open bags to avoid fermentation.

Seedling trees bear 5,000 to 10,000 small fruits per year in India. Superior grafted trees may yield as many as 30,000 fruits. The best cultivar in India, with fruits normally averaging 30 to the lb (66 to the kg), yields 175 lbs (77 kg) annually. Special cultural treatment increases both fruit size and yield.

World Production and Yield

The major production regions for Indian jujube are the arid and semi arid regions of India. From 1984 to 1995 with improved cultivars the production was 0.9 million tonnes on a land of 88,000 ha. The crop is also grown in Pakistan, Bangladesh and parts of Africa. Trees in northern India yield 80 to 200 kg of fresh fruit/tree/year when the trees are in their prime bearing age of 10–20 years.

Uses

The fruit is eaten raw, pickled or used in beverages. It is quite nutritious and rich in vitamin C. It is second only to guava and much higher than citrus or apples. In India, the ripe fruits are mostly consumed raw, but are sometimes stewed. Slightly underripe fruits are candied by a process of pricking, immersing in a salt solution. Ripe fruits are preserved by sun-drying and a powder is prepared for out-of-season purposes. It contains 20 to 30% sugar, up to 2.5% protein and 12.8% carbohydrates. Fruits are also eaten in other forms, such as dried, candied, pickled, as juice, or as ber butter. In Ethiopia, the fruits are used to stupefy fish. The leaves are readily eaten by camels, cattle and goats and are considered nutritious.

In India and Queensland, the flowers are rated as a minor source of nectar for honeybees. The honey is light and of fair flavor.

Ber timber is hard, strong, fine-grained, fine-textured, tough, durable, and reddish in colour. It has been used to line wells, to make legs for bedsteads, boat ribs, agricultural implements, tool handles, and other lathe-turned items. The branches are used as framework in house construction and the wood makes good charcoal with a heat content of almost 4,900 kcal per kg. In addition, this species is used as firewood in many areas. In tropical Africa, the flexible branches are wrapped as retaining bands around conical thatched roofs of huts, and are twined together to form thorny corral walls to retain livestock.

The fruits are applied on cuts and ulcers; are employed in pulmonary ailments and fevers; and, mixed with salt and chili peppers, are given in indigestion and biliousness. The dried ripe fruit is a mild laxative. The seeds are sedative and are taken, sometimes with buttermilk, to halt nausea, vomiting, and abdominal pains in pregnancy. They check diarrhea, and are poulticed on wounds. Mixed with oil, they are rubbed on rheumatic areas. The leaves are applied as poultices and are helpful in liver troubles, asthma and fever and, together with catechu, are administered when an astringent is needed, as on wounds. The bitter, astringent bark decoction is taken to halt diarrhea and dysentery and relieve gingivitis. The bark paste is applied on sores. The root is purgative. A root decoction is given as a febrifuge, taenicide and emmenagogue, and the powdered root is dusted on wounds. Juice of the root bark is said to alleviate gout and rheumatism. Strong doses of the bark or root may be toxic. An infusion of the flowers serves as an eye lotion.

The fatty-acid methyl ester of *Z. mauritiana* seed oil meets all of the major biodiesel requirements in the USA (ASTM D 6751-02, ASTM PS 121-99), Germany (DIN V 51606) and European Union (EN 14214). The average oil yield is 4.95 kg oil/tree or 1371 kg oil/hectare, and arid or semi-arid regions may be utilised due to its drought resistance.

Pests and Diseases

The greatest enemies of the jujube are fruit flies. Some cultivars are more susceptible than others, the flies preferring the largest, sweetest fruits, 100% of which may be attacked while on a neighbouring tree, bearing a smaller, less-sweet type, only 2% of the crop may be damaged. The larvae pupate in the soil and it has been found that treatment of the ground beneath the tree helps reduce the problem. Control is possible with regular and effective spraying of insecticide.

A leaf-eating caterpillar and the green slug caterpillar attack the foliage. Mites forms scale-like galls on twigs retarding growth and reducing the fruit crop.Lesser pests include a small caterpillar, *Meridarches scyrodes*, that bores into the fruit.

The tree is subject to shrouding by a parasitic vine . Powdery mildew causes defoliation and fruit-drop, but it can be adequately controlled. Lesser diseases are sooty mould, brown rot and leaf-spot. Leafspot results from infestation by *Cercospora* spp. and *Isariopsis indica* var. *zizyphi*. In 1973, a witches'-broom disease caused by a mycoplasma-like organism was found in jujube plants near Poona University. It proved to be transmitted by grafting or budding diseased scions onto healthy *Z. mauritiana* seedlings. Leaf rust, caused by *Phakopsora zizyphivulgaris*, ranges from mild to severe on all commercial cultivars in the Punjab.

In storage, the fruits may be spotted by the fungi. Fruit rots are caused by *Fusarium* spp., *Nigrospora oryzae*, *Epicoccum nigrum*, and *Glomerella cingulata*.

Phalsa (*Grewia Asiatica*)

Grewia asiatica (Phalsa or Falsa) is a species of *Grewia* native to southern Asia from Pakistan east to Cambodia, and widely cultivated in other tropical countries. *Grewia celtidifolia* was initially considered a mere variety of Phalsa, but is now recognized as a distinct species.

It is a shrub or small tree growing to 8 m tall. The leaves are broadly rounded, 5–18 cm long and broad, with a petiole 1-1.5 cm long. The flowers are produced in cymes of several together, the individual flowers about 2 cm diametre, yellow, with five large (12 mm) sepals and five smaller (4–5 mm) petals. The fruit is an edible drupe 5–12 mm diametre, purple to black when ripe.

Cultivation and Uses

It is extensively cultivated for its sweet and sour acidic fruits, which are sold in the market during summer months under the name Falsa. The pleasant sherbet or squash is prepared from the fruit pulp by mixing it with sugar and used as an astringent, stomachic and cooling agent.

These are said to be good for heart and blood disorders, fevers and diarrhoea. The fruit is also good for the troubles of throat. The root bark is used by Santhal tribals for rheumatism. The stem bark is said to be used in refining sugar, for making ropes and its infusion is used as a demulcent. The leaves are used as an application to pustular eruptions. The buds are also prescribed by some physicians.

It has become naturalised and locally invasive in Australia and the Philippines.

Jamun (*Syzygium Cuminii*)

Syzygium cumini, jambul, jambolan, jamblang, or jamun, is an evergreen tropical tree in the flowering plant family Myrtaceae. *Syzygium cumini* is native to Bangladesh, India, Nepal, Pakistan, Sri Lanka, Malaysia, the Philippines, and Indonesia. The name of the fruit is sometimes mistranslated as blackberry, which is a different fruit in an unrelated family.

The tree was introduced to Florida, USA in 1911 by the USDA, and is also now commonly grown in Suriname and Trinidad and Tobago. In Brazil, where it was introduced from India during Portuguese colonization, it has dispersed spontaneously in the wild in some places, as its fruits are eagerly sought by various native birds such as thrushes, tanagers and the Great Kiskadee.

This species is considered an invasive in Hawaii, USA. It is also illegal to grow, plant or transplant in Sanibel, Florida.

Description

A slow growing species, it can reach heights of up to 30 m and can live more than 100 years. Its dense foliage provides shade and is grown just for its ornamental value.

At the base of the tree, the bark is rough and dark grey, becoming lighter grey and smoother higher up. The wood is water resistant. Because of this it is used in railway sleepers and to install motors in wells. It is sometimes used to make cheap furniture and village dwellings though it is relatively hard to work on.

The leaves which are an aroma similar to turpentine, are pinkish when young, changing to a leathery, glossy dark green with a yellow midrib as they mature. The leaves are used as food for livestock, as they have good nutritional value.

Syzygium cumini trees start flowering from March to April. The flowers of are fragrant and small, about 5 mm in diametre. The fruits develop by May or June and resemble large berries. The fruit is oblong, ovoid, starts green and turns pink to shining crimson black as it matures. A variant of the tree produces white coloured fruit. The fruit has a combination of sweet, mildly sour and astringent flavour and tends to colour the tongue purple.

The seed is also used in various alternative healing systems like Ayurveda (to control diabetes, for example.), Unani and Chinese medicine for digestive ailments.

The pulp of the fruit, extracts from the bark and seeds is of great benefit when it comes to lowering of blood glucose level. Taking dried extract of the seeds orally, greatly reduces the blood sugar and glucosuria.

The leaves and bark are used for controlling blood pressure and gingivitis. Wine and vinegar are also made from the fruit. It has a high source in vitamin A and vitamin C.

Syzygium cumini has been spread overseas from India by Indian emigrants and at present is common in former tropical British colonies.

Cultural and Religious Significance

According to Hindu tradition, Rama subsisted on the fruit in the forest for 14 years during his exile from Ayodhya. Because of this, many Hindus regard *S. cumini* as a 'fruit of the gods,' especially in Gujarat, India, where it is known locally as jamboon. Lord Krishna has been described as having skin the colour of *S. cumini*. In Hindu mythology several protagonists have been described as having the colour of *S. cumini*.

Maharashtra

In Maharashtra, *S. cumini* (locally known as jâmbhû7 Marathi) leaves are used in marriage pandal decorations. There is famous Marathi song "Jambhul pikalya zada khali...". The seeds are used in herbal teas for diabetes

Andhra Pradesh

This tree is called Neredu((G0G!A) in Telugu. Besides the fruits, wood from Neredu tree is used in Andhra Pradesh to make bullock cart wheels and other agricultural equipment. Culturally, beautiful eyes are compared to this fruit. In the great epic of India Mahabharatha Sri Krishnas'[Lord Vishnu] body colour is compared to this fruit as well.

Tamil Nadu

There is a very famous legend that is associated with Auvaiyar (also Auvayar), a prominent female poets/ethicist/political activist of Sangam period (Tamil literature), and Naval Pazham(Jambu) in Tamil Nadu. Auvaiyar, believing to have achieved everything that is to be achieved, said to have been pondering over her retirement from Tamil literary work while resting under Naval Pazham tree. But she was met with and was wittily jousted by a disguised Lord Murugan (regarded as one of the guardian deities of Tamil language), who later revealed himself and made her realize that there is still a lot more to be done and learnt. Following this awakening, Auvaiyar is believed to have undertaken a fresh set of literary works, targeted at children.

Kerala

In Malayalam the tree is called *njaval* and its fruit are *njavalpazham*. The fruit is particularly plentiful in Kollam.

Karnataka- Kannada

In Kannada the tree is called *Nerale mara* and its fruit are *kempu Nerale Hannu* Nerale hannu is widely used by diabetes patients as it was thought to cure the same. The bears like this fruit. This tree is found everywhere in rural areas of Karnataka.

Karonda (*Carissa Congesta*)

Carissa carandas is a species of flowering shrub in the dogbane family, Apocynaceae. It produces berry-sized fruits that are commonly used as a condiment in Indian pickles and spices. It is a hardy, drought-tolerant plant that thrives well in a wide range of soils. Common names include *karonda, karamardaka* (Sanskrit), *vakkay* (Telugu), *kilaakkaai/* (Tamil). Other names less widely used include: *karau(n)da, karanda,* or *karamda.* It is called *kerenda* in Malaya, *karaunda* in Malaya and India; *Bengal currant* or *Christ's thorn* in South India; *nam phrom*, or *namdaeng* in Thailand; and *caramba, caranda, caraunda* and *perunkila* in the Philippines. In Assam it is called *Karja tenga.* In Bengali it is called as *Koromcha.*

***Figure:** The supposed varieties* congesta *and* paucinervia *actually refer to the related conkerberry (*C. spinarum*).*

Distribution

Carissa carandas grows naturally in the Himalayas at elevations of 300 to 1800 metres, in the Siwalik Hills, the Western Ghats and in Nepal and Afghanistan. It flourishes well in regions with high temperatures. In India it is grown on a limited scale in Rajasthan, Gujarat, Bihar and Uttar Pradesh. It is also grows in the Sri Lanka lowland rain forests.

Propagation

The plant is grown from seed sown in August and September. Vegetative propagation also is practiced in the form of budding and inarching. Cuttings may also succeed. The first monsoon shower is planting time. Plants raised from seed start bearing two years after planting. Flowering starts in March and in Northern India the fruit ripens from July to September.

Uses

***Figure:** Carissa carandas fruits ready for consumption*

The fruit is a rich source of iron, so it sometimes used in treatment of anaemia. It contains a fair amount of Vitamin C and therefore is an antiscorbutic. Mature fruit is harvested for pickles. It contains pectin and accordingly is a useful ingredient in jelly, jam, syrup and chutney. Ripe fruits exude a white latex when severed from the branch.

The roots of the plant are heavily branched, making it valuable for stabilizing eroding slopes.

Bael (*Aegle Marmelos*)

Aegle marmelos, commonly known as bael, Bengal quince, golden apple, stone apple, wood apple, bili, is a species of tree native to India. It is present throughout Southeast Asia as a naturalized species. The tree is considered to be sacred by Hindus. Its fruits are used in traditional medicine and as a food throughout its range.

Vernacular Names

The edible fruit tree is called "belada mara" and the religious tree "bilva" or "bilpathre" in Kannada. The fruits are known as *Belada Hannu* (edible variety), *Bilva* (sacred variety) in Kannada, "bela" in Oriya and *Maredu* (in Telugu).

Botanical Information

Bael is the only member of the monotypic genus *Aegle*. It is a mid-sized, slender, aromatic, armed, gum-bearing tree growing up to 18 metres tall. It has a leaf with three leaflets.

Ecology

Bael occurs in dry forests on hills and plains of northern, central and southern India, Pakistan, southern Nepal, Sri Lanka, Myanmar, Bangladesh, Vietnam, Laos, Cambodia and Thailand. It is cultivated throughout India, as well as in Sri Lanka, the northern Malay Peninsula, Java, the Philippines, and Fiji. It has a reputation in India for being able to grow in places that other trees cannot. It copes with a wide range of soil conditions (pH range 5-10), is tolerant of waterlogging and has an unusually wide temperature tolerance (from -7°C to 48°C). It requires a pronounced dry season to give fruit.

This tree is a larval foodplant for the following two Indian Swallowtail butterflies:

- The Lime Butterfly: *Papilio demoleus*
- The Common Mormon: *Papilio polytes*

Fruit

The bael fruit has a smooth, woody shell with a green, gray, or yellow peel. It takes about 11 months to ripen on the tree and can reach the size of a large grapefruit or pomelo, and some are even larger. The shell is so hard it must be cracked with a hammer or machete. The fibrous yellow pulp is very aromatic. It has been described as tasting of marmalade and smelling of roses. Boning (2006) indicates that the flavor is "sweet, aromatic and pleasant, although tangy and slightly astringent in some varieties. It resembles a marmalade made, in part, with citrus and, in part, with tamarind." Numerous hairy seeds are encapsulated in a slimy mucilage.

Uses

The fruit is eaten fresh or dried. If fresh, the juice is strained and sweetened to make a drink similar to lemonade. It can be made into sharbat (Hindi) or *Bela pana* or *bel pana*, a refreshing drink made of the pulp with water, sugar, and lime juice, mixed, left to stand a few hours, strained, and put on ice. One large bael fruit may yield five or six litres of sharbat.

If the fruit is to be dried, it is usually sliced and sun-dried. The hard leathery slices are then immersed in water.

The leaves and small shoots are eaten as salad greens.

In the system of Ayurveda this drug finds several and frequent therapeutic uses in different forms and recipes. They are prescribed in a number of diseases such as gastro intestinal diseases, piles, oedema, jaundice, vomiting, obesity, pediatric disorders, gynecological disorders, urinary complaints and as a rejuvenative. Besides the wide medicinal utility the plant and its certain parts (leaves and fruits) are of religious importance since the tree is regarded as one of the sacred trees of Indian heritage.

Aegeline (N-[2-hydroxy-2(4-methoxyphenyl) ethyl]-3-phenyl-2-propenamide) is a known constituent of the bael leaf and consumed as a dietary supplement for a variety of purposes.

Religious Significance-The Holy Bael

The fruit is also used in religious rituals. In Hinduism the tree is sacred. It is used in the worship of Shiva, who is said to favour the leaves. The tri-foliate form of leaves symbolize the trident that Shiva holds in his right hand. The fruits were used in place of coconuts before large-scale rail transportation became available. The fruit is said to resemble a skull with a white, bone-like outer shell and a soft inner part, and is sometimes called *seer phael* (head-fruit). However, it is quite likely that, the term 'Seer Phal' has coined from the Sanskrit term '*ShreePhal*, which again is a common name for this fruit. Many Hindus have bael trees in their gardens.

In the traditional Newari culture of Nepal, the bael tree is part of a fertility ritual for girls known as the *Bel baha*. Girls are "married" to the bael fruit and as long as the fruit is kept safe and never cracks the girl can never become widowed, even if her human husband dies. This was seen to be protection against the social disdain suffered by widows in the Newari community.

Medicinal Uses

Research has found the essential oil of the Bael tree to be effective against 21 types of bacteria. It is prescribed for smooth bowel movement to patients suffering from constipation and other gastrointestinal problems. Research also indicates that unripe Bael fruit is effective in combating giardia and rotavirus. While unripe Bael fruit did not show antimicrobial properties, it did inhibit bacteria adherence to and invasion of the gut (i.e. the ability to infect the gut).

Rambutan (*Nephelium Lappaccum*)

The rambutan (/ræmÈbuĐtYn/; taxonomic name: *Nephelium lappaceum*) is a medium-sized tropical tree in the family Sapindaceae. The fruit produced by the tree is also known as rambutan.

According to popular belief and the origin of its name, rambutan is native to Indonesia and Malaysia. The earliest record of rambutan trees show that they were cultivated by the Malayan jungle tribes around their temporary settlements, a practice followed to date. Rambutan trees grow naturally in Thailand, Vietnam, the Philippines, and elsewhere in Southeast Asia, although its precise natural distribution is unknown. It is closely related to several other edible tropical fruits including the lychee, longan, and mamoncillo. It is native to the Indonesian Archipelago, from where it spread westwards to Thailand, Burma, Sri Lanka and India; northwards to Vietnam, and the Philippines.

A species regularly sold in Costa Rican markets may be known as "wild" rambutan. Yellow in colour, it is smaller than the usual red variety. The flesh exposed when the outer skin is peeled off is sweet and sour, slightly grape-like and gummy to the taste. In Panama and Costa Rican Spanish, it is known as *mamón chino* ("Chinese Sucker") due to its Asian origin and the likeness of the edible part with *Melicoccus bijugatus*. The fruit has been successfully transplanted by grafting in Puerto Rico.

Rambutans are a non-climacteric fruit—that is, they ripen only on the tree.

Description

It is an evergreen tree growing to a height of 12–20 m. The leaves are alternate, 10–30 cm long, pinnate, with 3–11 leaflets, each leaflet 5–15 cm wide and 3–10 cm broad, with an entire margin. The flowers are small, 2.5–5 mm, apetalous, discoidal, and borne in erect terminal panicles 15–30 cm wide.

Rambutan trees are either male (producing only staminate flowers and, hence, produce no fruit), female (producing flowers that are only functionally female), or hermaphroditic (producing flowers that are female with a small percentage of male flowers).

The fruit is a round to oval single seeded berry, 3–6 cm (rarely to 8 cm) tall and 3–4 cm broad, borne in a loose pendant cluster of 10–20 together. The leathery skin is reddish (rarely orange or yellow), and covered with fleshy pliable spines, hence the name *rambutan*, derived from the Malay word *rambut*, which means *hairs*. The fruit flesh, which is actually the aril, is translucent, whitish or very pale pink, with a sweet, mildly acidic flavor very reminiscent of grapes.

The single seed is glossy brown, 1–1.3 cm, with a white basal scar. Soft and crunchy, the seeds may be cooked and eaten. Some folklore regards the seeds as poisonous, but tests of seed extracts reveal no toxicity to mice, even in doses up to 2500 mg/kg. The extract does show some analgesic activity, however. The peeled fruits can be cooked and eaten, first the grape-like aril, then the nutty seed, with no waste.

Production

It is a popular garden fruit tree and propagated commercially in small orchards. It is one of the best known fruits of Southeast Asia and is also widely cultivated elsewhere in the tropics including Africa, the Caribbean islands, Costa Rica, Honduras, Panama, India, Indonesia, the Philippines, and Sri Lanka; it is also produced in Ecuador where it is known as “achotillo” and on the island of Puerto Rico. Thailand is the largest producer from Surat Thani Province Thailand. Rambutan production is increasing in Australia and, in 1997, was one of the top three tropical fruits produced in Hawaii.

The fruit are usually sold fresh, used in making jams and jellies, or canned. Evergreen rambutan trees with their abundant coloured fruit make beautiful landscape specimens.

In India, Rambutan is imported from Thailand as well as grown in Pathanamthitta district of the southern state of Kerala.

Quality

The best quality rambutan is generally that which is harvested still attached to the branch. It is less susceptible to rot, damage, and pests, and remains fresh for a much longer time than rambutan that has been picked from the branch.

Another indicator of quality is the ease of detachment of the flesh from the seed. An easily detachable flesh normally will have bits of the woody seed coating. Thus, it is common Malay folklore to not eat too much rambutan when one has a cough.

Cultivation

Rambutan is adapted to warm tropical climates, around 22–30 °C, and is sensitive to temperatures below 10 °C. It is grown commercially within 12–15° of the equator. The tree grows well on heights up to 500 metres (1,600 ft) above sea-level and does best in deep soil; clay loam or sandy loam rich in organic matter, and thrive on hilly terrain as they require good drainage. Rambutan is propagated by grafting, air-layering, and budding; the latter is most common as trees grown from seed often produce sour fruit. Budded trees may fruit after 2–3 years with optimum production occurring after 8–10 years. Trees grown from seed bear after 5–6 years.

The aril is attached to the seed in some commercial cultivars, but “freestone” cultivars are available and in high demand. There is usually a single light brown seed, which is high in certain fats and oils (primarily oleic acid and arachidic acid) valuable to industry, and used in cooking and the manufacture of soap. Rambutan roots, bark, and leaves have various uses in medicine and in the production of dyes.

In some areas rambutan trees can bear fruit twice annually, once in late fall and early winter with a shorter season in late spring and early summer. In other areas like Costa Rica there is a single fruit season, with the start of the rainy season in April stimulating flowering and the fruit is usually ripe in August and September. The fragile nutritious fruit must ripen on the tree, then they are harvested over a four- to seven-week period.

The fresh fruit are easily bruised and have a limited shelf life. An average tree may produce 5,000–6,000 or more fruit (60–70 kg or 130–155 lb per tree). Yields begin at 1.2 tonnes per hectare (0.5 tons/acre) in young orchards and may reach 20 tonnes per hectare (8 tons per acre) on mature trees. In Hawaii, 24 of 38 cultivated hectares (60 of 95 acres) were harvested producing 120 tonnes of fruit in 1997. It has been suggested that yields could be increased via improved orchard management, including pollination, and by planting high yielding compact cultivars.

Most commercial cultivars are hermaphroditic; cultivars that produce only functionally female flowers require the presence of male trees. Male trees are seldom found as vegetative selection has favoured hermaphroditic clones that produce a high proportion of functionally female flowers and a much lower number of flowers that produce pollen. There are over 3,000 greenish-white flowers in male panicles, each with 5–7 anthers and a non-functional ovary. Male flowers have yellow nectaries and 5–7 stamens. There are about 500 greenish-yellow flowers in each hermaphroditic panicle. Each flower has six anthers, usually a bi-lobed stigma, and one ovule in each of its two sections (locules). The flowers are receptive for about one day but may persist if pollinators are excluded.

In Thailand, rambutan trees were first planted in Surat Thani in 1926 by the Chinese Malay K. Vong in Ban Na San. An annual rambutan fair is held during August harvest time.

In Malaysia, rambutan flowers from March to July and again between June and November, usually in response to rain following a dry period. Flowering periods differ for other localities. Most, but not all, flowers open early in the day. Up to 100 flowers in each female panicle may be open each day during peak bloom. Initial fruit set may approach 25% but a high level of abortion level contributes to a much lower level of production at harvest (1 to 3%). The fruit matures 15–18 weeks after flowering.

Rambutan cultivation in Sri Lanka mainly consists of small home gardens. Malwana, a village located in the Kelani River Valley, is popular for its rambutan orchards. Their production comes to market in the months of May, June and July, when it is very common to observe seasonal traders along the streets of Colombo. Sri Lanka also has some off-season rambutan production in the months of January and February in areas like Bibile, Medagama and Monaragala.

Both male and female flowers are faintly sweet scented and have functional nectaries at the ovary base. Female flowers produce 2–3 times more nectar than male flowers. Nectar sugar concentration ranges between 18% and 47% and is similar between the flower types. Rambutan is an important nectar source for bees in Malaysia.

Cross-pollination is a necessity because anther is absent in most functionally female flowers. Although apomixis may occur in some cultivars, research has shown that rambutan, like lychee, is dependent upon insects for pollination. In Malaysia, where only about one percent of the female flowers set fruit, research revealed that no fruit is set on bagged flowers while hand pollination resulted in 13 percent fruit set. These studies further suggest that pollinators may maintain a fidelity to either male or hermaphroditic flowers (trees), thus limiting pollination and fruit set under natural conditions where crossing between male and female flowers is required.

Cultivars

There are well over 200 cultivars developed from selected clones available throughout tropical Asia. Most of the cultivars are also selected for compact growth reaching a height of only 3–5 m for easier harvesting.

In Indonesia, there are 22 rambutan cultivars identified with good quality. Five of them are leading cultivars; Rambutan Rapiah, Rambutan Aceh Lebak Bulus, Rambutan Cimacan, Rambutan Binjai, and Rambutan Sinyonya.

In Nicaragua, a joint World Relief–European Union team distributed seedlings to organizations such as APAC (Ascociación Pueblos en Acción Comunitaria) in 2001 to more than 100 farmers. Some of these farmers saw the first production of rambutan from their trees in 2005–2006. The initiative is directed at the local market. However, it will be several years before growing practices and distribution are perfected. Other than seedling trees, the only known commercial cultivars in Nicaragua are R134, R162 and a Yellow Rambutan that lost its name when it was smuggled from Honduras in 2004–2005 (Frankie, J. A., Winrock International).

In the Philippines, two cultivars of rambutan are distinguished by its seed. The common rambutan seed and fruit are difficult to separate, while the Maharlika Rambutan fruit separates cleanly from the seed. The fruit taste and size of these two cultivars are identical. The Maharlika rambutan is far more popular and has a higher price. In India, one of the popular cultivars is Homegrown N 18, which is highly adaptable to Kerala, and is far superior to the existing cultivars.

Pollination

Aromatic rambutan flowers are highly attractive to many insects, especially bees. Flies (*Diptera*), bees (*Hymenoptera*) and ants (*Solenopsis*) are the main pollinators. Among the *Diptera*, *Lucilia* sp. is abundant and among the *Hymenoptera*, honey bees (*Apis dorsata* and *A. cerana*) and the stingless bee genus *Trigona* spp. are the major visitors. *Apis cerana* colonies foraging on rambutan flowers produce large quantities of honey. Bees foraging for nectar routinely contact the stigma of male flowers and gather significant quantities of the sticky pollen from male blossoms. Little pollen has been seen on bees foraging female flowers. Although male flowers open at 6 am, foraging by *A. cerana* is most intense between 7 am and 11 am, tapering off rather abruptly thereafter. In Thailand, *A. cerana* is the preferred species for small scale pollination of rambutan. Its hair is also helpful in pollination where pollen can be hooked on and transported to female flowers.

Durian (*Durio Zibethinus*)

The durian is the fruit of several tree species belonging to the genus *Durio* and the family Malvaceae (although some taxonomists place *Durio* in a distinct family, Durionaceae). Regarded by many people in southeast Asia as the "king of fruits", the durian is distinctive for its large size, strong odour, and formidable thorn-covered husk. The fruit can grow as large as 30 centimetres (12 in) long and 15 centimetres (6 in) in diametre, and it typically weighs one to three kilograms (2 to 7 lb). Its shape ranges from oblong to round, the colour of its husk green to brown, and its flesh pale yellow to red, depending on the species.

The edible flesh emits a distinctive odour that is strong and penetrating even when the husk is intact. Some people regard the durian as having a pleasantly sweet fragrance; others find the aroma overpowering and revolting. The smell evokes reactions from deep appreciation to intense disgust, and has been described variously as rotten onions, turpentine, and raw sewage. The persistence of its odour has led to the fruit's banishment from certain hotels and public transportation in Southeast Asia.

The durian, native to Southeast Asia, has been known to the Western world for about 600 years. The nineteenth-century British naturalist Alfred Russel Wallace famously described its flesh as "a rich custard highly flavoured with almonds". The flesh can be consumed at various stages of ripeness, and it is used to flavour a wide variety of savoury and sweet edibles in Southeast Asian cuisines. The seeds can also be eaten when cooked.

There are 30 recognised *Durio* species, at least nine of which produce edible fruit. *Durio zibethinus* is the only species available in the international market: other species are sold in their local regions. There are hundreds of durian cultivars; many consumers express preferences for specific cultivars, which fetch higher prices in the market.

Species

Durian trees are large, growing to 25–50 metres (80–164 ft) in height depending on the species. The leaves are evergreen, elliptic to oblong and 10–18 centimetres (4–7 in) long. The flowers are produced in three to thirty clusters together on large branches and directly on the trunk with each flower having a calyx (sepals) and five (rarely four or six) petals. Durian trees have one or two flowering and fruiting periods per year, although the timing varies depending on the species, cultivars, and localities. A typical durian tree can bear fruit after four or five years. The durian fruit can hang from any branch and matures roughly three months after pollination. The fruit can grow up to 30 centimetres (12 in) long and 15 centimetres (6 in) in diametre, and typically weighs one to three kilograms (2 to 7 lb). Its shape ranges from oblong to round, the colour of its husk green to brown, and its flesh pale-yellow to red, depending on the species. Among the thirty known species of *Durio*, nine of them have been identified as producing edible fruits: *D. zibethinus*, *D. dulcis*, *D. grandiflorus*, *D. graveolens*, *D. kutejensis*, *D. macrantha*, *D. oxleyanus*, and *D. testudinarum*. There are many species for which the fruit has never been collected or properly examined, however, so other species with edible fruit may exist. The durian is somewhat similar in appearance to the jackfruit, an unrelated species.

The name *durian* comes from the Indonesian and Malaysian word *duri* (thorn) together with the suffix *-an* (for building a noun in Indonesia and Malaysia). *D. zibethinus* is the only species commercially cultivated on a large scale and available outside of its native region. Since this species is open-pollinated, it shows considerable diversity in fruit colour and odour, size of flesh and seed, and tree phenology. In the species name, *zibethinus* refers to the Indian civet, *Viverra zibetha*. There is disagreement regarding whether this name, bestowed by Linnaeus, refers to civets being so fond of the durian that the fruit was used as bait to entrap them, or to the durian smelling like the civet.

Durian flowers are large and feathery with copious nectar, and give off a heavy, sour, and buttery odour. These features are typical of flowers pollinated by certain species of bats that eat nectar and pollen. According to research conducted in Malaysia in the 1970s, durians were pollinated almost exclusively by cave fruit bats (*Eonycteris spelaea*); however, a 1996 study indicated two species, *D. grandiflorus* and *D. oblongus*, were pollinated by spiderhunters (Nectariniidae) and another species, *D. kutejensis*, was pollinated by giant honey bees and birds as well as bats.

Cultivars

Figure: *Different cultivars of durian often have distinct colours. D101 (right) has rich yellow flesh, clearly distinguishable from another variety (left).*

Over the centuries, numerous durian cultivars, propagated by vegetative clones, have arisen in southeast Asia. They used to be grown with mixed results from seeds of trees bearing superior quality fruit, but now are propagated by layering, marcotting, or more commonly, by grafting, including bud, veneer, wedge, whip or U-grafting onto seedlings of randomly selected rootstocks. Different cultivars may be distinguished to some extent by variations in the fruit shape, such as the shape of the spines. Durian consumers express preferences for specific cultivars, which fetch higher prices in the market.

Most cultivars have a common name and a code number starting with "D". Mon thong is the most commercially sought after for its thick, full-bodied creamy and mild sweet tasting flesh with relatively moderate smell emitted and smaller seeds, while Chanee is the best in terms of its resistance to infection by *Phytophthora palmivora*. Kan Yao is somewhat less common, but prized for its longer window of time when it is both sweet and odorless at the same time. Among all the cultivars in Thailand, five are currently in large-scale commercial cultivation: Chanee, Mon Thong, Kan Yao, Ruang, and Kradum. There are more than 100 registered cultivars since 1920's in Malaysia and up to 193 cultivar by 1992, and many superior cultivars have been identified through competitions held at the annual Malaysian Agriculture, Horticulture, and Agrotourism Show. In Vietnam, the same process has been achieved through competitions held by the Southern Fruit Research Institute. A recently popular variety is, Cat Mountain King.

By 2007, Songpol Somsri, a Thai government scientist, had crossbred more than ninety varieties of durian to create Chantaburi No. 1, a cultivar without the characteristic odour. Another hybrid, Chantaburi No. 3, develops the odour about three days after the fruit is picked, which enables an odourless transport yet satisfies consumers who prefer the pungent odour. On 22 May 2012, two other cultivars from Thailand that also lack the usual odour, Long Laplae and Lin Laplae, were presented to the public by Yothin Samutkhiri, governor of Uttaradit province from where these cultivars were developed locally, while he announced the dates for the yearly durian fair of Laplae District, and the name giver to both cultivars.

Cultivation and Availability

The durian is native to Brunei, Indonesia and Malaysia. There is some debate as to whether the durian is native to the Philippines, particularly the Davao region in the island of Mindanao, or was introduced. The durian is grown in other areas with a similar climate; it is strictly tropical and stops growing when mean daily temperatures drop below 22 °C (72 °F).

The centre of ecological diversity for durians is the island of Borneo, where the fruit of the edible species of *Durio* including *D. zibethinus*, *D. dulcis*, *D. graveolens*, *D. kutejensis*, *D. oxleyanus* and *D. testudinarum* is sold in local markets. In Brunei, *D. zibethinus* is not grown because consumers prefer other species such as *D. graveolens*, *D. kutejensis* and *D. oxleyanus*. These species are commonly distributed in Brunei, and together with other species like *D. testudinarum* and *D. dulcis*, represent rich genetic diversity.

Although the durian is not native to Thailand, the country is currently one of the major exporters of durians, growing 781,000 tonnes of the world's total harvest of 1,400,000 tonnes in 1999, 111,000 tonnes of which it exported to Taiwan, Hong Kong, Malaysia, Singapore and Canada. Malaysia and Indonesia follow, both producing about 265,000 tonnes each. Of this, Malaysia exported 35,000 tonnes in 1999. Chantaburi in Thailand each year holds the World Durian Festival in early May. This single province is responsible for half of the durian production of Thailand. In the Philippines, the centre of durian production is the Davao Region. The Kadayawan Festival is an annual celebration featuring the durian in Davao City. Other places where durian farms are located include Cambodia, Laos, Vietnam, Myanmar, Sri Lanka, India, the West Indies, Florida, Hawaii, Papua New Guinea, the Polynesian Islands, Madagascar, southern China (Hainan Island), northern Australia, and Singapore.

***Figure:** Durians being sold in mesh bags out of a freezer in a California market*

Durian was introduced into Australia in the early 1960s and clonal material was first introduced in 1975. Over thirty clones of *D. zibethinus* and six *Durio* species have been subsequently introduced into Australia. China is the major importer, purchasing 65,000 tonnes in 1999, followed by Singapore with 40,000 tonnes and Taiwan with 5,000 tonnes. In the same year, the United States imported 2,000 tonnes, mostly frozen, and the European Community imported 500 tonnes.

The durian is a seasonal fruit, unlike some other non-seasonal tropical fruits such as the papaya, which are available throughout the year. In Peninsular Malaysia and Singapore, the season for durians is typically from June to August, which coincides with that of the mangosteen. Prices of durians are relatively high as compared with other fruits. For example, in Singapore, the strong demand for high quality cultivars such as the D24, *Sultan*, and *Mao Shan Wang* has resulted in typical retail prices of between S$8 to S$15 (US$5 to US$10) per kilogram of whole fruit. With an average weight of about 1.5 kilograms (3.3 lb), a durian fruit would therefore cost about S$12 to

S$22 (US$8 to US$15). The edible portion of the fruit, known as the aril and usually referred to as the "flesh" or "pulp", only accounts for about 15–30% of the mass of the entire fruit. Many consumers in Singapore are nevertheless quite willing to spend up to around S$75 (US$50) on a single purchase of about half a dozen of the favoured fruit to be shared by family members.

In-season durians can be found in mainstream Japanese supermarkets, while in the West they are sold mainly by Asian markets.

Flavour and Odour

The unusual flavour and odour of the fruit have prompted many people to express diverse and passionate views ranging from deep appreciation to intense disgust. Writing in 1856, the British naturalist Alfred Russel Wallace provides a much-quoted description of the flavour of the durian:

The five cells are silky-white within, and are filled with a mass of firm, cream-coloured pulp, containing about three seeds each. This pulp is the edible part, and its consistence and flavour are indescribable. A rich custard highly flavoured with almonds gives the best general idea of it, but there are occasional wafts of flavour that call to mind cream-cheese, onion-sauce, sherry-wine, and other incongruous dishes. Then there is a rich glutinous smoothness in the pulp which nothing else possesses, but which adds to its delicacy.

It is neither acid nor sweet nor juicy; yet it wants neither of these qualities, for it is in itself perfect. It produces no nausea or other bad effect, and the more you eat of it the less you feel inclined to stop. In fact, to eat Durians is a new sensation worth a voyage to the East to experience. ... as producing a food of the most exquisite flavour it is unsurpassed.

Wallace described himself as being at first reluctant to try it because of the aroma, "but in Borneo I found a ripe fruit on the ground, and, eating it out of doors, I at once became a confirmed Durian eater." He cited one traveller from 1599: "it is of such an excellent taste that it surpasses in flavour all other fruits of the world, according to those who have tasted it." He cites another writer: "To those not used to it, it seems at first to smell like rotten onions, but immediately after they have tasted it they prefer it to all other food. The natives give it honourable titles, exalt it, and make verses on it." Despite having tried many foods that are arguably more eccentric, Andrew Zimmern, host of Bizarre Foods, was unable to finish a durian upon sampling it, due to his intolerance of its strong taste.

While Wallace cautions that "the smell of the ripe fruit is certainly at first disagreeable", later descriptions by westerners are more graphic. British novelist Anthony Burgess writes that eating durian is "like eating sweet raspberry blancmange in the lavatory". Chef Andrew Zimmern compares the taste to "completely rotten, mushy onions". Anthony Bourdain, a lover of durian, relates his encounter with the fruit thus: "Its taste can only be described as...indescribable, something you will either love or despise. ...Your breath will smell as if you'd been French-kissing your dead grandmother." Travel and food writer Richard Sterling says:

> *... its odor is best described as pig-shit, turpentine and onions, garnished with a gym sock. It can be smelled from yards away. Despite its great local popularity, the raw fruit is forbidden from some establishments such as hotels, subways and airports, including public transportation in Southeast Asia.*

Other comparisons have been made with the civet, sewage, stale vomit, skunk spray and used surgical swabs. The wide range of descriptions for the odour of durian may have a great deal to do with the variability of durian odour itself. Durians from different species or clones can have significantly different aromas; for example, red durian (*D. dulcis*) has a deep caramel flavour with a turpentine odour while red-fleshed durian (*D. graveolens*) emits a fragrance of roasted almonds. Among the varieties of *D. zibethinus*, Thai varieties are sweeter in flavour and less odorous than Malay ones.

The degree of ripeness has an effect on the flavour as well. Three scientific analyses of the composition of durian aroma – from 1972, 1980, and 1995 – each found a mix of volatile compounds including esters, ketones, and different sulphur compounds, with no agreement on which may be primarily responsible for the distinctive odour.

This strong odour can be detected half a mile away by animals, thus luring them. In addition, the fruit is extremely appetising to a variety of animals, including squirrels, mouse deer, pigs, orangutan, elephants, and even carnivorous tigers.

While some of these animals eat the fruit and dispose of the seed under the parent plant, others swallow the seed with the fruit and then transport it some distance before excreting, with the seed being dispersed as a result. The thorny, armoured covering of the fruit discourages smaller animals; larger animals are more likely to transport the seeds far from the parent tree.

Aonla (*Emblica Officinalis*)

Phyllanthus emblica (syn. *Emblica officinalis*), the Indian gooseberry, or Dhatrik (in Maithili), or amla from Sanskrit amalika is a deciduous tree of the family Phyllanthaceae. It is known for its edible fruit of the same name.

Plant Anatomy and Harvesting

The tree is small to medium in size, reaching 8 to 18 m in height, with a crooked trunk and spreading branches. The branchlets are glabrous or finely pubescent, 10–20 cm long, usually deciduous; the leaves are simple, subsessile and closely set along branchlets, light green, resembling pinnate leaves. The flowers are greenish-yellow. The fruit is nearly spherical, light greenish yellow, quite smooth and hard on appearance, with six vertical stripes or furrows.

Ripening in autumn, the berries are harvested by hand after climbing to upper branches bearing the fruits. The taste of Indian gooseberry is sour, bitter and astringent, and it is quite fibrous. In India, it is common to eat gooseberries steeped in salt water and turmeric to make the sour fruits palatable. It is also used to straighten hair.

Amla, aonla or Indian gooseberry (Emblica officinalis) is an indigenous fruit to the Indian subcontinent. Owing to hardy nature, suitability to various waste lands, high productivity/unit area (15-20tons/ha), nutritive and therapeutic value, amla is becoming more and more commercially important with every passing year.

Amla fruits are a very rich source of vitamin C having an ascorbic acid content varying from 0.9 to 1.3 per cent. This is the second highest among all the cultivated fruits. This fruit is highly valued among indigenous medicines. It is acrid, cooling, refrigerant, diuretic and laxative. Dried fruits have been reported to be useful in haemorrhages, diarrhea, dysentery, anaemia, jaundice, dyspepsia and cough. Trifala and Chyavanprash are well known indigenous medicines in Ayurvedic system using amla fruits.

Foliage and Fruits of Amla

Murabba (preserve) is prepared from amla fruits by keeping them in sugar syryp. This is a very popular amla preparation in India and hundreds of tons are sold every year. The fruits are also used to prepare hairwash and hair oil which are said to be beneficial for growth and appearance of hair. Besides fruits, leaves, bark and even seeds are being used for various purposes. Its cultivation is common in India, particularly in Pratapgarh, Rai Bareilly, Varanasi, Jaunpur, Sultanpur,

Kanpur, Agra and Mathura districts of Uttar Pradesh. Its intensive plantation is being done in the salt affected areas of the state of Uttar Pradesh, including ravenous areas in Agra, Mathura, Etawah, Fatehpur and semiarid tract of Bundelkhand. Amla cultivation is also spreading rapidly in the semiarid regions of Maharashtra, Gujarat, Rajasthan, Andhra Pradesh, Karnataka, Tamil Nadu, Aravali ranges in Haryana and Kandi area in Punjab and Himachal Pradesh.

Climate and Soil

Amla is a subtropical plant and prefers dry subtropical climate. Even mild frost during the winter months can be injurious to tree. Though it can bear temperature upto 46 C during summer months, but the temperature should not be high at the time of flowering. It affects fruit setting and may sometime lead to complete unfruitfulness, if there are hot and dry winds too. Ample humidity is essential for initiation of fruit growth of dormant fruitlets during July –August. Dry spells result in heavy dropping and delay in initiation of fruit growth.

Since amla is a hardy plant, it can be successfully grown in variable soil conditions. The deep root system, reduced foliage, dormancy of fertilized fruitlets (April- June) makes amla an ideal plant for arid and semiarid conditions. Aonla can be cultivated in marginal soils –slightly acidic to saline/sodic (pH 6.5-9.5) conditions. Heavy soils or high water table areas are not suited for cultivation.

Varieties

There are 3 main varieties of amla viz., Banarasi, Francis (Hathijhool) and Chakaiya. These varieties have their own merits and demerits. Banarasi, an early-maturing amla, is a shy-bearing, prone to heavy dropping of fruits with poor shelf –life. Francis suffers from severe incidence of fruit necrosis. Chakaiya fruits are fibrous, smaller in size and also have a tendency to bear heavy crop in alternate years. Other varieties identified and released for commercial cultivation during the recent years are:

Fruits of Banarasi Amla: Kanchan (NA4): A seedling selection from Chakaiya, it is heavy and regular bearer (7.7female flowers/ branchlet), with medium-sized fruits, having higher fibre content. It is preferred by industriefor pulp extraction and manufacturing of various products. This has been adopted very well in the semiarid regions of Gujarat and Maharashtra.

NA6: A seedling selection from Chakaiya, it is prolific and heavy bearer (10.8female flowers / branchlet). It is ideal for preserve and candy, owing to low fibre content.

NA7: A seedling selection of Francis, it is precocious, prolific and regular-bearer (9.7female flowers/ branchlet). This is an ideal variety for preparation of products and has a great promise. Besides above varieties, Anand 1, Anand 2 and Anand 3 have been selected as promising strains at the Gujarat Agricultural University.

Wild Himalayan amla: There also exists a quite distinct strain of amla, which grows wild in the mid-hill regions of the Western Himalayas. This amla bears smaller fruits and is a very heavy cropper. A very good quality of this strain is that it is much cold hardy in comparison to the commercial amla. The wild Himalayan amla can therefore be successfully used to extend amla cultivation to colder areas.

A separate chapter on this strain under the heading "Wild Himalayan amla" has been given here.

Propagation and Rootstock

Amla has are raised through seeds as well as inarching. However, there is prolonged juvenility and wide variability in seedlings. On the other hand only limited number of scion shoots is available for inarching owing to upright tree habit. It can also be propagated through patch/ modified ring budding in north India during mid May to September with 60-100%sucess. Besides, Veneer grafting also has successfully been attempted. Considering the efficiency and requirement of single bud, budding is an ideal method of propagation.

Six months to one year-old seedlings obtained from ordinary seedlings amla trees are being used as rootstock. Mature amla fruits are collected during January-February and their seeds are extracted after drying. Seeds are sown in raised beds April onwards and these are transplanted in separate bed for subsequent budding.

Cultivation

Planting: Grafted or budded amla plants are planted 7-10 m apart during July–August or February. Pits of 1-1.25 m size are dug 2 months prior to planting. In each pit 3-4 baskets of well rotten farmyard manure and 1 kg neem cake or 500 g bone meal are mixed with soil and filled. In sodic soils, 5-8 kg gypsum along with 20kg sand is incorporated. Filled pits are irrigated thoroughly if there is no rain. Hedge row planting is also being tried keeping line–to–line distance of 8m, while plant-to-plant distance is reduced to 4-5m. Under adverse soil conditions, it is advisable to grow the seedlings directly in the field pits or raise these in suitable containers and transplant at permanent site, and perform budding (in situ) subsequently. Amla scion shoots can be safely stored for 5-7 days with ample success.

Amla is a Heavy Bearer Tree

Since self incompatibility appears to be a problem with amla varieties, 2 varieties in alternate rows need to be planted.

Pruning

Amla plant should be encouraged to develop a medium headed tree. The main branches should be allowed to appear at a height of 0.75-1m above the ground level. Plants should be trained to modified central leader system. Two to four branches with wide crotch angle, appearing in the opposite directions should be encouraged in early years. The unwanted branches in are pinched off during March-April. In the subsequent years, 4-6 branches should be allowed to develop. Regular pruning of a bearing aonla tree is not required. As per growth habit, shedding of all determinate shoots encourages new growth in coming season. However, dead, infested, broken, weak or overlapping branches should be removed regularly.

Manuring and Fertilization

The dose of manures and fertilizers depends upon soil fertility, age of plant and production. A dose of 10 kg farmyard manure, 100 g N, 50 g P and 100 g K should be given to one years old plants of aonla. This dose should be increased yearly up to 10 years and thereafter a constant dose should be given. Full dose of farmyard manure and P and half of N and K should be given in tree basin during January-February. The remaining half should be applied in August. In sodic soils, 100-500 g of B and zinc sulphate should also be incorporated along with fertilizers as per tree age and vigour.

Irrigation

Established amla orchards in general do not require irrigation particularly in normal soils. No irrigation is required during rainy and winter season. However, irrigation at 15-20 days interval is desirable in dry summer particularly during early years of orchard establishment under wasteland conditions. Brackish water should not be used for irrigation. In the bearing plantation, first irrigation should be given just after manure and fertilizer application (January/ February). Irrigation should be avoided during flowering (mid-March-mid April) period. Irrigation at 10-15 days intervals should be given particularly in the salt affected soils. Basin system of irrigation is well suited for amla. The initial information obtained with drip irrigation has shown promising response. Alternate drip irrigation with 60 CPE is appropriate with water saving of 40-45%. In water scarcity areas, pitcher irrigation can also be successfully utilized.

Mulching

Mulching with organic wastes is very effective tool for establishment of amla orchards in sodic and ravenous areas. Paddy straw, sugarcane trash and farmyard manure have shown better response. Mulching with organic wastes over a number of years shall be helpful in improving the organic matter content, infiltration rate, and restricting the upward movement or soluble salts and thus escaping their toxicity menace in salt-affected soils.

Intercropping

Amla being a deep rooted, deciduous tree with sparse foliages, is an ideal plant amicable for 2 or 3 tier cropping system. Fruits, vegetables, flowers and a few medicinal and aromatic plants are well suited for intercropping in aonla orchards

Harvesting and Post Harvest Management

Change in seed colour from creamy white to brown is an indication of fruit maturity. Fully developed fruits are harvested. Delay in harvesting results in heavy dropping of fruits particularly in Banarasi and Francis. It also adversely affects the following years bearing. Individual fruits are plucked by climbing on the tree with the help of pegged bamboo or ladder. Harvesting should be done in early or in the late hours of the day. A budded/grafted aonla tree starts bearing third year onwards after planting, whereas a seedlings tree may take 6-8 years. Vegetatively propagated plants attain full bearing within 10-12 years and may continue to bear for 60-75 years of age under well managed conditions. An aonla tree may bear 1-300 kg tree, giving 15-20 tons/ha.

Amla fruits are graded into 3 grades. Large sized, sound fruits are mostly utilized for preserve and candy; small sized for chyavanprash and trifala and blemished fruits for powder and shampoo making. Aonla fruits can be stored for 6-9 days at ambient temperature. However, with a salt solution it can be stored up to 75 days.

Physiological Disorder

Necrosis, a physiological disorder, has been observed in aonla fruits. Francis variety is highly susceptible followed by Banarasi incidence initiates with browning of mesocarp which extends towards the epicarp resulting into brownish black appearance of flesh.

Jackfruit (*Artocarpus Heterophyllus*)

The jackfruit (*Artocarpus heterophyllus*, also known as jack tree, jakfruit, or sometimes simply jack or jak) is a species of tree in the

Artocarpus genus of the mulberry family (*Moraceae*). It is native to parts of South and Southeast Asia, and is believed to have originated in the southwestern rain forests of India, in present-day Kerala, coastal Karnataka and Maharashtra. The jackfruit tree is well suited to tropical lowlands, and its fruit is the largest tree-borne fruit, reaching as much as 80 pounds (36 kg) in weight, 36 inches (90 cm) in length, and 20 inches (50 cm) in diametre.

The jackfruit tree is a widely cultivated and popular food item in tropical regions of India, Bangladesh, Nepal, Sri Lanka, Cambodia, Vietnam, Thailand, Malaysia, Indonesia, and the Philippines. Jackfruit is also found across Africa (e.g., in Cameroon, Uganda, Tanzania, Madagascar, and Mauritius), as well as throughout Brazil and in Caribbean nations such as Jamaica. Jackfruit is the national fruit of Bangladesh.

Cultivation and Ecology

***Figure:** Jackfruit garden in Bangladesh*

The jackfruit has played a significant role in Indian agriculture for centuries. Archeological findings in India have revealed that jackfruit was cultivated in India 3000 to 6000 years ago. It is also widely cultivated in southeast Asia.

Thailand and Vietnam are major producers of jackfruit, a lot of which are cut, prepared and canned in a sugary syrup (or frozen in bags/ boxes without syrup), and exported overseas, frequently to North America and Europe.

In other areas, the jackfruit is considered an invasive species as in Brazil's Tijuca Forest National Park in Rio de Janeiro. The Tijuca is mostly an artificial secondary forest, whose planting began during the mid-19th century, and jackfruit trees have been a part of the park's flora since its founding. Recently, the species has expanded excessively; its fruits, which naturally fall to the ground and open, are eagerly eaten by small mammals such as the common marmoset and coati. The seeds are dispersed by these animals, which allows the jackfruit to compete for space with native tree species. Additionally, as the marmoset and coati also prey opportunistically on bird's eggs and nestlings, the supply of jackfruit as a ready source of food has allowed them to expand their populations, to the detriment of the local bird populations. Between 2002 and 2007, 55,662 jackfruit saplings were destroyed in the Tijuca Forest area in a deliberate culling effort by the park's management.

Aroma

Jackfruit are known for having a distinct aroma. In a study using five jackfruit cultivars, the main jackfruit volatile compounds that were detected are: ethyl isovalerate, 3-methylbutyl acetate, 1-butanol, propyl isovalerate, isobutyl isovalerate, 2-methylbutanol, and butyl isovalerate. These compounds were consistently present in all the five cultivars studied, suggesting that these esters and alcohols contributed to the sweet and fruity aroma of jackfruit.

Fruit

The flesh of the jackfruit is starchy and fibrous and is a source of dietary fibre. The flavor is comparable to a combination of apple, pineapple, mango and banana. Varieties are distinguished according to characteristics of the fruit's flesh. In Brazil, three varieties are recognized: *jaca-dura*, or the "hard" variety, which has a firm flesh and the largest fruits that can weigh between 15 and 40 kilograms each, *jaca-mole*, or the "soft" variety, which bears smaller fruits with a softer and sweeter flesh, and *jaca-manteiga*, or the "butter" variety, which bears sweet fruits whose flesh has a consistency intermediate between the "hard" and "soft" varieties. In Indochina, there are 2 varieties, being the "hard" version (more crunchy, drier and less sweet but fleshier), and the "soft" version (more soft, moister, much sweeter with a darker gold-colour flesh than the hard variety).

In Kerala, two varieties of jackfruit predominate: *varikka* and *koozha* . *Varikka* has a slightly hard inner flesh when ripe, while the inner flesh of the ripe *koozha* fruit is very soft and almost dissolving. A sweet preparation called *chakka varattiyathu* (jackfruit jam) is made by seasoning pieces of *varikka* fruit flesh in jaggery, which can be preserved and used for many months. Huge jackfruits up to four feet in length with a corresponding girth are sometimes seen in Kerala.

In West Bengal the two varieties are called *khaja kathal* and *moja kathal*. The fruits are either eaten alone or as a side to rice / roti / chira / muri. Sometimes the juice is extracted and either drunk straight or as a side with muri. The extract is sometimes condensed into rubbery delectables and eaten as candies. The seeds are either boiled or roasted and eaten with salt and hot chillies. They are also used to make spicy side-dishes with rice or roti.

In Mangalore, Karnataka, the varieties are called *bakke* and *imba*. The pulp of the *imba* jackfruit is ground and made into a paste, then spread over a mat and allowed to dry in the sun to create a natural chewy candy.

The young fruit is called *polos* in Sri Lanka and *idichakka* or *idianchakka* in Kerala.

In Indochina, jackfruit is a frequent ingredient in sweets and desserts. In Vietnam, jackfruit is used to make jackfruit Chè (*chè* is a sweet dessert soup, similar to the Chinese derivative, *bubur chacha*). The Vietnamese also use jackfruit puree as part of pastry fillings, or as a topping on Xôi ngÍt (sweet version of sticky rice portions).

Ripe jackfruit is naturally sweet with subtle flavoring. It can be used to make a variety of dishes, including custards, cakes, *halo-halo* and more. In India, when the Jackfruit is in season, an ice cream chain store called "Naturals" carries Jackfruit flavored ice cream.

Ripe jackfruit arils are sometimes seeded, fried or freeze-dried and sold as jackfruit chips.

The seeds from ripe fruits are edible, are said to have a milky, sweet taste, and may be boiled, baked or roasted. When roasted the flavor of the seeds is comparable to chestnuts. Seeds are used as snacks either by boiling or fire roasted, also used to make desserts. For making the traditional breakfast dish in southern India: idlis, the fruit is used along with rice as an ingredient and jackfruit leaves are used as a wrapping for steaming. Jackfruit dosas can be prepared by grinding jackfruit flesh along with the batter.

Seeds

In general, the seeds are gathered from the ripe fruit, sun-dried, then stored for use in rainy season in many parts of South Indian states. They are extracted from fully matured fruits and washed in water to remove the slimy part. Seeds should be stored immediately in closed polythene bags for one or two days to prevent them from drying out. Germination is improved by soaking seeds in clean water for 24 hours. During transplanting, sow seeds in line, 30 cm apart, in a nursery bed filled with 70% soil mixed with 30% organic matter. The seedbed should be shaded partially from direct sunlight in order to protect emerging seedlings.

Boiled Jackfruit seed is also edible. Seasoned with nothing more than salt, this snack is very popular in Java.

Wood

The wood of the tree is used for the production of musical instruments. In Indonesia, hardwood from the trunk is carved out to form the barrels of drums used in the *gamelan*, and in the Philippines its soft wood is made into the body of the *kutiyapi*, a type of boat lute. It is also used to make the body of the Indian string instrument *veena* and the drums *mridangam*, *thimila* and *kanjira*; the golden, yellow timber with good grains is used for building furniture and house construction in India. The ornate wooden plank called *avani palaka* made of the wood of jackfruit tree is used as the priest's seat during Hindu ceremonies in Kerala. In Vietnam, jackfruit wood is prized for the making of Buddhist statuaries in temples.

Jackfruit wood is widely used in the manufacture of furniture, doors and windows, and in roof construction. The heartwood is used by Buddhist forest monastics in Southeast Asia as a dye, giving the robes of the monks in those traditions their distinctive light-brown colour.

Commercial Availability

Outside of its countries of origin, fresh jackfruit can be found at Asian food markets, especially in the Philippines, Thailand, Vietnam, Malaysia Cambodia and Bangladesh. It is also extensively cultivated in the Brazilian coastal region, where it is sold in local markets. It is available canned in sugary syrup, or frozen, already prepared and cut. Dried jackfruit chips are produced by various manufacturers. In northern Australia, particularly in Darwin, jackfruit can be found at outdoor produce markets during the dry season. Outside of countries where it is grown, jackfruit can be obtained year-round both canned or dried. It has a ripening season in Asia of late spring to late summer.

There are established jackfruit industries in Sri Lanka and Vietnam, where the fruit is processed into products such as flour, noodles, papad and ice cream. It is also canned and sold as a vegetable for export.

Soil, Nutrients and Propagation

A mulch is a layer of material applied to the surface of an area of soil. Its purpose is any or all of the following:

- to conserve moisture
- to improve the fertility and health of the soil
- to reduce weed growth
- to enhance the visual appeal of the area

A mulch is usually but not exclusively organic in nature. It may be permanent (e.g. plastic sheeting) or temporary (e.g. bark chips). It may be applied to bare soil, or around existing plants. Mulches of manure or compost will be incorporated naturally into the soil by the activity of worms and other organisms. The process is used both in commercial crop production and in gardening, and when applied correctly can dramatically improve soil productivity.

Uses

Many materials are used as mulches, which are used to retain soil moisture, regulate soil temperature, suppress weed growth, and for aesthetics. They are applied to the soil surface, around trees, paths, flower beds, to prevent soil erosion on slopes, and in production areas for flower and vegetable crops. Mulch layers are normally two inches or more deep when applied. They are applied at varies times of the year depending on the purpose.

Towards the beginning of the growing season mulches serve initially to warm the soil by helping it retain heat which is lost during the night. This allows early seeding and transplanting of certain crops, and encourages faster growth. As the season progresses, mulch stabilizes the soil temperature and moisture, and prevents the growing of weeds from seeds. In temperate climates, the effect of mulch is dependent upon the time of year they are applied and when applied in fall and winter, are used to delay the growth of perennial plants in the spring or prevent growth in winter during warm spells, which limits freeze thaw damage.

The effect of mulch upon soil moisture content is complex. Mulch forms a layer between the soil and the atmosphere which prevents sunlight from reaching the soil surface, thus reducing evaporation. However, mulch can also prevent water from reaching the soil by absorbing or blocking water from light rains.

In order to maximise the benefits of mulch, while minimizing its negative influences, it is often applied in late spring/early summer when soil temperatures have risen sufficiently, but soil moisture content is still relatively high. However, permanent mulch is also widely used and valued for its simplicity, as popularized by author Ruth Stout, who said, "My way is simply to keep a thick mulch of any vegetable matter that rots on both sides of my vegetable and flower garden all year long. As it decays and enriches the soils, I add more."

Plastic mulch used in large-scale commercial production is laid down with a tractor-drawn or standalone layer of plastic mulch. This is usually part of a sophisticated mechanical process, where raised beds are formed, plastic is rolled out on top, and seedlings are transplanted through it. Drip irrigation is often required, with drip tape laid under the plastic, as plastic mulch is impermeable to water.

In home gardens and smaller farming operations, mulches are usually spread by hand around emerged plants. (On plots with existing mulch, the mulch is pulled away from the seedbed before planting, and restored after the seedlings have emerged.) For materials like straw and hay, a shredder may be used to chop up the material. Organic mulches are usually piled quite high, six inches (152 mm) or more, and settle over the season.

Materials

Materials used as mulches vary and depend on a number of factors. Use takes into consideration availability, cost, appearance, the effect it has on the soil—including chemical reactions and pH, durability, combustibility, rate of decomposition, how clean it is—some can contain weed seeds or plant pathogens.

A variety of materials are used as mulch:

- Organic residues: grass clippings, leaves, hay, straw, kitchen scraps comfrey, shredded bark, whole bark nuggets, sawdust, shells, woodchips, shredded newspaper, cardboard, wool, animal manure, etc. Many of these materials also act as a direct composting system, such as the mulched clippings of a mulching lawn mower, or other organics applied as sheet composting.
- Compost: fully composted materials are used to avoid possible phytotoxicity problems. Materials that are free of seeds are ideally used, to prevent weeds introduced by the mulch.
- Rubber mulch: made from recycled tire rubber.
- Plastic mulch: crops grow through slits or holes in thin plastic sheeting. This method is predominant in large-scale vegetable growing, with millions of acres cultivated under plastic mulch worldwide each year (disposal of plastic mulch is cited as an environmental problem).
- Rock and gravel can also be used as a mulch. In cooler climates the heat retained by rocks may extend the growing season.

In some areas of the United States, such as central Pennsylvania and northern California, mulch is often referred to as "tanbark", even by manufacturers and distributors. In these areas, the word "mulch" is used specifically to refer to very fine tanbark or peat moss.

Organic Mulches

Organic mulches decay over time and are temporary. The way a particular organic mulch decomposes and reacts to wetting by rain and dew affects its usefulness.

Some mulches such as straw, peat, sawdust and other wood products may for a while negatively affect plant growth because of their wide carbon to nitrogen ratio, because bacteria and fungi that decompose the materials remove nitrogen from the surrounding soil for growth. However, whether this effect has any practical impact on gardens is disputed by researchers and the experience of gardeners. Organic mulches can mat down, forming a barrier that blocks water and air flow between the soil and the atmosphere. Vertically applied organic mulches can wick water from the soil to the surface, which can dry out the soil. Mulch made with wood can contain or feed termites, so care must be taken about not placing mulch too close to houses or building that can be damaged by those insects. Some mulch manufacturers recommend putting mulch several inches away from buildings.

Commonly available organic mulches include:

Leaves

- *Leaves* from deciduous trees, which drop their foliage in the autumn/fall. They tend to be dry and blow around in the wind, so are often chopped or shredded before application. As they decompose they adhere to each other but also allow water and moisture to seep down to the soil surface. Thick layers of entire leaves, especially of maples and oaks, can form a soggy mat in winter and spring which can impede the new growth lawn grass and other plants. Dry leaves are used as winter mulches to protect plants from freezing and thawing in areas with cold winters, they are normally removed during spring.

Grass Clippings

- *Grass clippings*, from mowed lawns are sometimes collected and used elsewhere as mulch. Grass clippings are dense and tend to mat down, so are mixed with tree leaves or rough compost to provide aeration and to facilitate their decomposition without smelly putrefaction. Rotting fresh grass clippings can damage plants; their rotting often produces a damaging buildup of trapped heat. Grass clippings are often dried thoroughly before application, which mediates against rapid decomposition and excessive heat generation. Fresh green grass clippings are relatively high in nitrate content, and when used as a mulch, much of the nitrate is returned to the soil, but the routine removal of grass clippings from the lawn results in nitrogen deficiency for the lawn.

Peat Moss

- *Peat moss*, or sphagnum peat, is long lasting and packaged, making it convenient and popular as a mulch. When wetted and dried, it can form a dense crust that does not allow water to soak in. When dry it can also burn, producing a smoldering fire. It is sometimes mixed with pine needles to produce a mulch that is friable. It can also lower the pH of the soil surface, making it useful as a mulch under acid loving plants.

However peat bogs are a valuable wildlife habitat, and peat is also one of the largest stores of carbon (in Britain, out of a total estimated 9952 million tonnes of carbon in British vegetation and soils, 6948 million tonnes carbon are estimated to be in Scottish, mostly peatland, soils), so gardeners who wish to protect the environment will choose more sustainable alternatives.

Wood Chips

- *Wood chips* are a byproduct of the pruning of trees by arborists, utilities and parks; they are used to dispose of bulky waste. Tree branches and large stems are rather coarse after chipping and tend to be used as a mulch at least three inches thick. The chips are used to conserve soil moisture, moderate soil temperature and suppress weed growth. The decay of freshly produced chips from recently living woody plants, consumes nitrate; this is often off set with a light application of a high-nitrate fertilizer. Wood chips are most often used under trees and shrubs. When used around soft stemmed plants, an unmulched zone is left around the plant stems to prevent stem rot or other possible diseases. They are often used to mulch trails, because they are readily produced with little additional cost outside of the normal disposal cost of tree maintenance. Wood chips come in various colours.

Woodchip Mulch

- *Woodchip mulch* is a byproduct of reprocessing used (untreated) timber (usually packaging pallets), to dispose of wood waste by creating woodchip mulch. The chips are used to conserve soil moisture, moderate soil temperature and suppress weed growth. Woodchip mulch is often used under trees, shrubs or large planting areas and can last much longer than arborist mulch. In addition, many consider woodchip mulch to be visually appealing, as it comes in various colours. Woodchips can also be reprocessed into playground woodchip to be used as an impact-attenuating playground surfacing.

Bark Chips

- *Bark chips* of various grades are produced from the outer corky bark layer of timber trees. Sizes vary from thin shredded strands to large coarse blocks. The finer types are very attractive but have a large exposed surface area that leads to quicker decay. Layers two or three inches deep are usually used, bark is relativity inert and its decay does not demand soil nitrates. Bark chips are also available in various colours.

Straw Mulch / Field Hay / Salt Hay

- *Straw mulch* or *field hay* or *salt hay* are lightweight and normally sold in compressed bales. They have an unkempt look and are used in vegetable gardens and as a winter covering. They are biodegradable and neutral in pH. They have good moisture retention and weed controlling properties but also are more likely to be contaminated with weed seeds. Salt hay is less likely to have weed seeds than field hay. Straw mulch is also available in various colours.

Cardboard / Newspaper

- *Cardboard* or *newspaper* can be used as mulches. These are best used as a base layer upon which a heavier mulch such as compost is placed to prevent the lighter cardboard/newspaper layer from blowing away. By incorporating a layer of cardboard/newspaper into a mulch, the quantity of heavier mulch can be reduced, whilst improving the weed suppressant and moisture retaining properties of the mulch. However, additional labour is expended when planting through a mulch containing a cardboard/newspaper layer, as holes must be cut for each plant. Sowing seed through mulches containing a cardboard/newspaper layer is impractical. Application of newspaper mulch in windy weather can be facilitated by briefly pre-soaking the newspaper in water to increase its weight.

Coloured Mulch

Some organic mulches are coloured red, brown, black, and other colours. Isopropanolamine, specifically 1-Amino-2-propanol or DOW™ monoisopropanolamine, may be used as a pigment dispersant and colour fastener in these mulches. Types of mulch which can be dyed include: wood chips, bark chips (barkdust) and pine straw. Coloured mulch is made by dyeing the mulch in a water-based solution of colourant and chemical binder. When coloured mulch first entered the market, most formulas were suspected to contain toxic, heavy metals and other contaminates. Today, "current investigations indicate that mulch colourants pose no threat to people, pets or the environment.

The dyes currently used by the mulch and soil industry are similar to those used in the cosmetic and other manufacturing industries (i.e., iron oxide)," as stated by the Mulch and Soil Council. Coloured mulch can be applied anywhere non-coloured mulch is used (such as large bedded areas or around plants) and features many of the same gardening benefits as traditional mulch, such as improving soil productivity and retaining moisture. As mulch decomposes, just as with non-coloured mulch, more mulch may need to be added to continue providing benefits to the soil and plants. However, if mulch is faded, spraying dye to previously spread mulch in order to restore colour is an option.

Anaerobic (Sour) Mulch

Mulch normally smells like freshly cut wood, but sometimes develops a toxicity that causes it to smell like vinegar, ammonia, sulfur or silage. This happens when material with ample nitrogen content is not rotated often enough and it forms pockets of increased decomposition. When this occurs, the process may become anaerobic and produce these phytotoxic materials in small quantities. Once exposed to the air, the process quickly reverts to an aerobic process, but these toxic materials may be present for a period of time. If the mulch is placed around plants before the toxicity has had a chance to dissipate, then the plants could very likely be damaged or killed depending on their hardiness. Plants that are predominantly low to the ground or freshly planted are the most susceptible, and the phytotoxicity may prevent germination of some seeds.

If sour mulch is applied and there is plant kill, the best thing to do is to water the mulch heavily. Water dissipates the chemicals faster and refreshes the plants. Removing the offending mulch may have little effect, because by the time plant kill is noticed, most of the toxicity is already dissipated. While testing after plant kill will not likely turn up anything, a simple pH check may reveal high acidity, in the range of 3.8 to 5.6 instead of the normal range of 6.0 to 7.2. Finally, placing a bit of the offending mulch around another plant to check for plant kill will verify if the toxicity has departed. If the new plant is also killed, then sour mulch is probably not the problem.

Groundcovers (Living Mulches)

Groundcovers are plants which grow close to the ground, under the main crop, to slow the development of weeds and provide other benefits of mulch. They are usually fast-growing plants that continue growing with the main crops. By contrast, cover crops are incorporated into the soil or killed with herbicides. However, live mulches also may need to be mechanically or chemically killed eventually to prevent competition with the main crop.

Some groundcovers can perform additional roles in the garden such as nitrogen fixation in the case of clovers, dynamic accumulation of nutrients from the subsoil in the case of creeping comfrey (*Symphytum ibericum*), and even food production in the case of *Rubus tricolour.*

On-site Mulch Production

Owing to the great bulk of mulch which is often required on a site, it is often impractical and expensive to source and import sufficient mulch materials. An alternative to importing mulch materials is to grow them on site in a "mulch garden" - an area of the site dedicated entirely to the production of mulch which is then transferred to the growing area. Mulch gardens should be sited as close as possible to the growing area so as to facilitate transfer of mulch materials.

Mulching (Composting) Over Unwanted Plants

Sufficient mulch over plants will destroy them, and may be more advantageous than using herbicide, cutting, mowing, pulling, raking, or tilling. The higher the temperature that this "mulch" is composted, the quicker the reduction of undesirable materials. "Undesirable materials" may include living seed, plant "trash", as well as pathogens such as from animal feces, urine (e.g. hantavirus), fleas, lice, ticks, etc.

In some ways this improves the soil by attracting and feeding earthworms, and adding humus. Earthworms "till" the soil, and their feces are among the best fertilizers and soil conditioners.

Urine may be toxic to plants if applied to growing areas undiluted.

Use of Bioregulators in Fruit Production

Plant bioregulators are unique agrochemicals in that they must be absorbed by the plant tissue and transported to a reaction site before the desired response can be achieved. Most plant bioregulators are applied as aqueous sprays to a variety of plant surfaces. The total surface area and, in particular, its surface chemistry and morphology are key factors in determining the bioregulator dose that will be retained and, hence, available for penetration. Essentially, all aerial plant organs are covered with a thin cuticle or cuticular membrane (CM), a noncellular, lipoidal covering that serves as the prime barrier to penetration of foliar applied chemicals.

The CM is composed of a cutin matrix of esterified long chain hydroxylated fatty acids. The outer morphological surface is usually covered with difficult-to-wet epicuticular waxes, and a well-defined layer of highly oriented cuticular waxes is found embedded in the cutin matrix. While the surface chemistry and morphology (particularly roughness) determine the wetting properties and degree of spray retention, the structure and chemistry of the CM influence transcuticular penetration. Spray additives, like surface active agents, may alter the physio-chemical properties of the spray solution, thus improving wetting, spray retention and coverage.

Penetration, however, is affected mainly by the nature (e.g., polarity, charge, and solubility), concentration and formulation of the penetrant, and by temperature and humidity. Specific surfactants may also enhance penetration not only by increasing the interaction between the penetrant and CM, but also by plasticizing the waxes and/or cutin polymer. The physical nature and composition of the chemical deposit formed on the plant surface after evaporation of the aqueous phase of the spray droplets appear to have a marked effect on penetration from the apparently dry deposits.

The production of most economically important fruit crops involves the use of one or more plant bioregulating chemicals. Some fruit crops could not be produced without some critical PBR tool; others would be significantly more expensive, available over a shorter season, or would be less attractive because of deficiencies in fruit quality. Increasingly, PBRs are being recognized as important tools in pest and disease management.

However, despite their immense usefulness in horticulture, low hazard to human health, and great potential for providing "soft" options in pest and disease control, the future of PBR usage in fruit production is uncertain. When any one commercial practice or product is judged to be unsustainable or unsafe it places all PBR products in jeopardy. Thus, it is very important that future research on PBR chemicals focuses as much on specificity, mode of action and degradation as on efficacy in a commercial application.

Bioregulators have been used for many years to alter the behaviour of fruit or fruit trees for the economic benefit of the fruit grower. Control of vegetative vigor, stimulation of flowering, regulation of crop load, reduction of fruit drop, and delay or stimulation of fruit maturity and ripening are important examples of processes in fruit and fruit trees that can be regulated with exogenous applications of bioregulators. New candidate bioregulators with possible benefits for fruit growers are continually being made available by private industry.

In addition, we are continuing to find new uses for bioregulator products that have been made available for other uses. In our research programme we seek to develop information that can help Washington tree fruit growers to benefit economically from the use of bioregulators in the orchard and nursery.

Propagation of Fruit Plants

Fruit tree propagation is usually carried out vegetatively (non-sexually) by grafting or budding a desired variety onto a suitable rootstock.

Perennial plants can be propagated either by sexual or vegetative means. Sexual reproduction begins when a male germ cell (pollen) from one flower fertilises a female germ cell (ovule, incipient seed) of the same species, initiating the development of a fruit containing seeds. Each seed, when germinated, can grow to become a new specimen tree.

However, the new tree inherits characteristics of both its parents, and it will not grow 'true' to the variety of either parent from which it came. That is, it will be a fresh individual with an unpredictable combination of characteristics of its own.

Although this is desirable in terms of producing novel combinations from the richness of the gene pool of the two parent plants (such sexual recombination is the source of new cultivars), only rarely will the resulting new fruit tree be directly useful or attractive to the tastes of humankind. Most new plants will have characteristics that lie somewhere between those of the two parents.

Therefore, from the orchard grower or gardener's point of view, it is preferable to propagate fruit cultivars vegetatively in order to ensure reliability.

This involves taking a cutting (or scion) of wood from a desirable parent tree which is then grown on to produce a new plant or 'clone' of the original. In effect this means that the original Bramley apple tree, for example, was a successful variety grown from a pip, but that every Bramley since then has been propagated by taking cuttings of living matter from that tree, or one of its descendants.

Methods

The simplest method of propagating a tree vegetatively is rooting or taking cuttings. A cutting (usually a piece of stem of the parent plant) is cut off and stuck into soil. Artificial rooting hormones are sometimes used to improve chances of success. If the cutting does not die from rot-inducing fungi or desiccation first, roots grow from the buried portion of the cutting to become a new complete plant. However although this works well for some plants (such as figs and olives), for most fruit tree cultivars this method has much too low a success rate to be commercially viable. Root cuttings (pieces of root cut off and induced to grow a new trunk) are also not used to propagate fruit trees, although this method is successful with some herbaceous plants.

A refinement on rooting is layering. This is rooting a piece of a wood that is still attached to its parent and continues to receive nourishment from it. The new plant is severed only after it has successfully grown roots. Layering is the technique most used for propagation of clonal apple rootstocks. The most common method of propagating fruit trees, suitable for nearly all species, is grafting onto rootstocks. This in essence involves physically joining part of a shoot of a hybrid cultivar onto the roots of a different but closely related species or cultivar, so that the two parts grow together as one plant.

The process of joining the two varieties must ensure maximum contact between the cambium (the layer just below the bark) of each, so that they grow together successfully. Grafting is a preferred method because it not only propagates a new plant of the desired hybrid cultivar, it usually also confers extra advantages as a result of the characteristics of the rootstocks (or stocks), which are selected for characteristics such as their vigour of growth, hardiness and soil tolerance, as well as compatibility with the desired variety that will form the aerial part of the plant (called the scion). For example, grape rootstocks descended from North American grapes allow European grapes to be grown in areas infested with *Phylloxera*, a soil-dwelling insect that attacks and kills European grapes when grown on their own roots. Two of the most common grafting techniques are 'whip and tongue', carried out in spring as the sap rises, and 'budding', which is performed around the end of summer.

Bud Grafting

1. Cut a slice of bud and bark from the parent tree.
2. Cut a similar sliver off the rootstock, making a little lip at the base to slot the scion into.

3. Join the two together and bind.
4. In time, the scion bud will grow into a shoot, which will develop into the desired tree.

Whip and Tongue Grafting

1. Make a sloping cut in the rootstock with a 'tongue' pointing upwards.
2. Make a matching cut in the scion wood with a 'tongue' pointing downwards.
3. Join the two, ensuring maximum contact of the vascular cambium layers. Bind with raffia or polythene tape or wind around with a 5mm wide strip of elastic band (this is particularly successful because it keeps pressure on the cambium layers to be joined and eventually falls away without cutting into the bark as the tree grows) and seal with grafting wax.

Bibliography

Arguello D. S.: *Production, Post-harvest Management and Exportation of Tropical Fruits in Costa Rica*, Wageningen, Technical Centre for Agricultural and Rural Cooperation, 1992.

Arora, Chandralekha : *Advances in Horticulture Science Research*, Oxford Book Company, Delhi, 2009.

Barbeau G.: *Tropical Fruits in Nicaragua*, Managua, Nicaragua Ministerio de Desarrollo Agropecuario, Agraria, 1990.

Bhattacharjee, Supriya Kumar : *Amenity Horticulture, Biotechnology and Post-Harvest Technology*, Pointer, Delhi, 2006.

Chadha K. L. and Pareek O. P.: *Advances in Horticulture: Fruit Crops,* New Delhi, Malhotra Publishing House, 1993.

Crucefix D.: *Avocado Variety Selection for Export Development,* Roseau, CARDI, 1996.

Currah L. and Proctor F. J.: *Onions in Tropical Regions*, Kent, Natural Resources Institute, 1990.

Dhatt A. S. and Singh Z.: *Propagation and Rootstocks of Citrus*, New Delhi, Malhotra Publishers, 1993.

Doijode S. D.: *Seed Germination in Fruits*, New Delhi, Malhotra Publishers, 1993.

Featherly H. I.: *Taxonomic Terminology of the Higher Plants*, USA, Iowa State College Press, 1954.

Gibbs, D.R. : *Chemotaxonomy of Flowering Plants*, McGill-Queen's University Press, Montreal, 1974.

Glowinski Louis: *The Complete Book of Fruit Growing in Australia*, Australia, Lothian Publishing Company Pty. Ltd., 1991.

Gunjate R. T. and Tawde A. B.: *Propagation and Rootstocks of Mango*, New Delhi, Malhotra, 1993.

Harlan, Jack R.: *The Living Fields: Our Agricultural Heritage*, Cambridge: Cambridge University Press, 1998.

Herklots G. A. C.: *Vegetables in South East Asia*, London, George Allen & Unwin Ltd., 1972.

Jeffers P.: *Evaluation of Four Onion Varieties in Montserrat*, Plymouth, CARDI, 1992.

Jerram, M.R.K. : *A Text-Book on Forest Management : In Agriculture, Horticulture and Forestry*, Asiatic Pub, Delhi, 2006.

Jugale, V.B. : *Horticulture Economy of Maharashtra*, Shruti, Delhi, 2004.

Malins A.: *Postharvest Handling of Pineapple and Mango*, Port of Spain, Trinidad and Tobago, 1992.

Matsuoka H.: *Cultivation of Panicum Genetic Resources for Evaluation of Characteristics*, Tokyo, JICA, 1997.

Mitra S.: *Postharvest Physiology and Storage of Tropical and Subtropical Fruits*, Oxon, CABI, 1997.

Morton Julia F.: *Fruits of Warm Climates*, Miami, Julia F. Morton Publisher, 1987.

Oldham P.: *Cost of Production of Major Tree Crops in Dominica*, Roseau, Ministry of Agriculture, 1991.

Peter, K.V. : *Basics of Horticulture*, New India, Delhi, 2009.

Pilgrim R.: *Post Harvest Handling of Minor Exotics*, St. George's, Grenada, 1996.

Prasad, Surendra : *Agro's Dictionary of Horticulture*, Agrobios, Delhi, 2003.

Sharma, V.K. : *Advances in Horticulture : Strategies, Production, Plant Protection and Value Addition*, Deep and Deep, Delhi, 2004.

Singh H. P. and Chadha K. L.: *Genetic Resources of Citrus*, New Delhi, Malhotra Publishing House, 1993.

Singh, D.K. : *Elements of Horticulture*, Agrotech, Delhi, 2005.

Singh, H.P. : *Challenges of Climate Change - Indian Horticulture*, Westville Publishing House, Delhi, 2010.

Singh, S.P. : *Advances in Horticulture and Forestry*, Scientific, Delhi, 2001.

Thavarasook C.: *Proceedings of the Mungbean Meeting 90, Chiang Mai, Thailand,* Bangkok, Tropical Agricultural Research Centre, 1991.

Weiss, R.F. : *Herbal Medicine*, Beaconsfield Arcanum, Beaconsfield, 1991 .

Whitwell A.: *Dominica Orchard Crop Management and Research Project: Pest and Disease Management,* United Kingdom, Natural Resources Institute, 1991.

Index

L

M

N

O

P

R

S

T

W